Erosion and sedimentation

Erosion and sedimentation

PIERRE Y. JULIEN

CAMBRIDGE
UNIVERSITY PRESS

PUBLISHED BY THE PRESS SYNDICATE OF THE UNIVERSITY OF CAMBRIDGE
The Pitt Building, Trumpington Street, Cambridge CB2 1RP

CAMBRIDGE UNIVERSITY PRESS
The Edinburgh Building, Cambridge CB2 2RU, United Kingdom
40 West 20th Street, New York, NY 10011-4211, USA
10 Stamford Road, Oakleigh, Melbourne 3166, Australia

First published 1994
First paperback edition 1998

Printed in the United States of America

Library of Congress Cataloging-in-Publication Data is available.

A catalog record for this book is available from the British Library.

ISBN   0-521-44237-0 hardback
ISBN   0-521-63639-6 paperback

To my family, Helga and Patrick

Contents

Preface

Erosion and sedimentation processes have fascinated generations of researchers, and yet significant contributions to this rapidly evolving scientific field are still to be made. The state of the art in erosion and sedimentation can be assessed only through a careful examination of both theoretical developments and practical engineering technology. This book has been prepared for graduate students and visiting scholars actively pursuing scientific research and for practitioners keeping up with recent technological developments. The prerequisites simply include a basic knowledge of undergraduate fluid mechanics and a fundamental understanding of partial differential equations.

As a physical science, the topic is deeply rooted in the realm of fluid mechanics, on the basis of which the mechanics of hydrodynamic forces applied on single particles, suspensions, and hyperconcentrations can best be understood. This text is not a voluminous encyclopedia; rather, it scrutinizes carefully selected methods that meet pedagogical objectives underlining theory and applications. The material can be covered within a regular forty-five-hour course at most academic institutions.

The chapters of this book contain a variety of exercises, general problems, computer problems, examples, and case studies. Each illustrates a specific aspect of the profession, from theoretical derivations through exercises, to practical solutions to real problems through the analysis of case studies. Most problems can be solved with a few algebraic equations; others require the use of a computer. Problems marked with a double asterisk are considered very important; those with a single asterisk or no asterisk are less important. Throughout, a solid diamond (♦) is used to denote equations of particular significance.

Recent technological developments in engineering promote the use of computers for qualitative analyses of erosion, transport, and sedimentation. The computer problems offer students the opportunity to develop skills for formulating physical problems in numerical form in order to be solved on digital computers. No specific computer code or language is required. Instead of using existing software packages, this textbook promotes student creativity and originality in developing computerized tools to solve physical problems of erosion and sedimentation numerically.

I am grateful to Professor Marcel Frenette and Professor Hunter Rouse, who deeply influenced my academic development, and Professor D. B. Simons and Professor E. V. Richardson for their encouragement and support during my course on erosion and sedimentation at Colorado State University. This book benefited greatly from numerous suggestions offered by graduate students. Jenifer Davis diligently typed successive drafts of the manuscript with the assistance of Hélène Michel, and Jean Parent prepared the figures. Finally, it has been a great pleasure to collaborate with Florence Padgett, Mary Racine, and the Cambridge University Press production staff.

Notation

Symbols

a, a_x	acceleration
a	reference elevation
A	surface area
A_T, A_U, A_G, A_B	gross, upland, gully, and bank erosion
A_t	drainage area
c	wave celerity
c_{Bd}, c_{cl}	coefficients
C	Chézy coefficient
C	sediment concentration
Co	particle shape factor
C_v, C_s, C_{ppm}, $C_{mg/l}$	sediment concentration
C_t, $C_{\forall}$, C_f	time-, spatial-, and flux-averaged concentration
C_D, C_E	drag and expansion coefficients
$\hat{C}$	cropping-management factor
d_5, d_{50}	grain size
d_*	dimensionless particle diameter
D	molecular diffusion coefficient
e	void ratio
e_B	Bagnold coefficient
E	specific energy
f	Darcy–Weisbach friction factor
F	force
Fr	Froude number
g	gravitational acceleration
G	specific gravity
Gr	gradation coefficient
h	flow depth

h_n	normal flow depth
h_c	critical flow depth
H	Bernoulli sum
He	Hedstrom number
i	rainfall intensity
I	universal soil-loss equation rainfall intensity
I_1, I_2	Einstein integrals
k_s	surface roughness
K	consolidation coefficient
$\hat{K}$	soil erodibility factor
K_d	dispersion coefficient
l	liter
l, L	lengths
l_m	mixing length
L_b, L_s, L_t	bedload, suspended load, total load
$\hat{L}$	field length factor
m, M	mass
M, N	particle stability coefficients
M_D', M_D''	moments
n	Manning coefficient
$\vec{n}$	vector normal to a surface
p	pressure
p_0	porosity
P	wetted perimeter
$\hat{P}$	conservation practice factor
q	unit discharge
q_b, q_s, q_t	unit sediment discharge
Q	total discharge
Q_b, Q_s, Q_t	sediment discharge
r	radial coordinate
R	radius of a sphere
$\hat{R}$	rainfall erosity factor
Re	Reynolds number
Re_*	grain shear Reynolds number
Re_d	densimetric Reynolds number
Re_p	particle Reynolds number
Re_B	Bingham Reynolds number
Ro	Rouse number
R_h	hydraulic radius

Sh	Shen–Hung parameter
S_0, S_f, S_w	bed, friction, and water surface slopes
$\hat{S}$	slope steepness factor
SF	particle stability factor
S_{DR}	sediment–delivery ratio
t, T	time
t_s	sampling
t_v	mixing time scales
T	sediment transport parameter
T°	temperature
T_E	trap efficiency
T_R	life expectancy
T_w	wave period
T_c	consolidation time
u, v	velocity
u_*	shear velocity
V	depth-averaged flow velocity
$\forall$	volume
W	channel width
x, y, z	coordinates
X_C	settling distance
X_D	total rate of energy dissipation
X_r	runoff length
Y	sediment yield
Z	dependent variable

Greek symbols

α_e	energy correction factor
β	particle motion angle
β_m	momentum correction factor
β_s	ratio of sediment to momentum exchange coefficient
γ	specific weight
Γ	circulation
δ	laminar sublayer thickness
Δ	dune height
Δp_i	sediment size fraction
ϵ	turbulent mixing coefficient
ϵ_m	eddy viscosity

κ	von Kármán constant
η_0, η_1	particle stability number
λ	streamline deviation angle
Λ	dune wavelength
μ	dynamic viscosity
ν	kinematic viscosity
ϕ	angle of repose
ρ	mass density
Π	dimensionless parameter
ω	fall velocity
$\vec{\omega}$	vorticity
Ω_g	gravitation potential
Ω_e	elastic energy
θ	angular coordinate
Θ	angle
ϑ	mixing stability parameter
σ	stress components
σ_g	gradation coefficient
σ_t	mixing width
τ	shear stress
τ_*	Shields parameter
τ_y, τ_d	yield and dispersive stresses
Φ	potential function
Ψ	stream functions
χ	rate of work done per unit mass
χ_D	dissipation function
ζ	turbulent-dispersive parameter
$\odot$	translation
$\ominus$	linear deformation
$\oslash$	angular deformation
$\otimes$	rotation

Superscripts and diacriticals

$\hat{v}$	fluctuating parameter
$\bar{v}$	average value
v^+	time fluctuation
τ'	grain resistance
τ''	form resistance
$\tilde{V}$	integrated value

Subscripts

a_x, a_y	x, y components
h_n	normal flow depth
h_c	critical flow depth
τ_0, τ_b, τ_s	boundary shear stress, at the bed, and on the side slope
τ_{yx}	shear stress in the x direction from gradient in the y direction
q_{bv}, q_{bw}	unit sediment discharge by volume and by weight
X_d, X_t, X_v	lengths for dispersion, transversal, and vertical mixing
L_b, L_s	bedload, suspended load
L_m, L_u	measured, unmeasured load
L_w, L_{bm}	washload, bed material load
h_d, V_d	densimetric values

1

Introduction to erosion and sedimentation

Erosion and sedimentation embody the processes of erosion, transportation, and deposition of solid particles, often called sediments. These natural processes have been active throughout geological time and have shaped the present landscape of our world. Today, erosion, transport, and sedimentation can cause severe engineering and environmental problems.

Human activities exert a profound influence on erosion. Under some circumstances, the erosion rate is 100 times greater than the normal, or geological, erosion rate. The erodibility of natural materials is enhanced by disturbances to the soil structure due to plowing and tillage. The protective canopy is weakened by grubbing, cutting, or burning of existing vegetation. Besides producing harmful sediment, erosion may cause serious on-site damage to agricultural land by reducing the fertility and productivity of soils. Runoff conditions on land surfaces and the hydraulic characteristics of flow in channels are exacerbated by improvements in surface drainage and by alterations in the characteristics of natural channels such as meander cutoffs.

Severe erosion can occur during the construction of roads and highways when protective vegetation is removed and steeply sloping cuts and fills are left unprotected. Such erosion can cause local scour problems along with serious sedimentation downstream. Approximately 85% of the 571,000 bridges in the United States are built over waterways. The majority of these bridges span rivers and streams that are continuously adjusting their beds and banks. Bridges on more active streams can be expected to experience scour problems as a result of stream realignment. Local scour at bridge piers and erosion of abutments are the most common causes of bridge failure during floods.

Mining operations may introduce large volumes of sediment directly into natural streams. Mine dumps and spoil banks often continue to erode by natural rainfall for many years after mining operations have ceased. For example, some drainage and flood problems in the Sacramento Valley,

1

California, as well as problems of construction and maintenance of navigation channels, can be traced directly to mining activities that took place more than a century ago. Gravel stream mining can cause severe channel instabilities such as upstream headcutting, which may trigger a complex response of the fluvial system.

Stream and river control works may have a serious local influence on channel erosion. Channel straightening, which increases slope and flow velocity, may initiate channel erosion. If the bed of a main stream is lowered, the beds of tributary streams are also lowered. In many instances, such bed degradation is beneficial because it restores the flood-carrying capacity of channels. However, where parent materials are eroded by a new set of hydraulic conditions, degradation may proceed far beyond the original bed levels and actually initiate an entirely new erosion cycle on the upstream watershed.

Sediment transport affects water quality and its suitability for human consumption or use in various enterprises. Numerous industries cannot tolerate even the smallest amount of sediment in the water that is necessary for certain manufacturing processes, and the public pays a large price for the removal of sediments from the water it consumes every day.

Dam construction influences channel stability in two ways. It traps the incoming sediment, and it changes the natural flow and sediment load downstream. As a net result, degradation occurs below dams and aggradation might increase the risk of flooding upstream of the reservoir. The loss of storage capacity in reservoirs in the United States amounts to a monetary loss of $100 million annually. Severe problems of abrasion of turbines, dredging, and stream instability and possible failure are often associated with reservoir and dam construction. Damage can be observed downstream from dam failure sites.

Sediment not only is the major water pollutant, but also serves as a catalyst, carrier, and storage agent of other forms of pollution. Sediment alone degrades water quality for municipal supply, recreation, industrial consumption and cooling, hydroelectric facilities, and aquatic life. In addition, chemicals and waste are assimilated onto and into sediment particles. Ion exchange occurs between solutes and sediments. Thus, sediment particles have become a source of increased concern as carriers and storage agents of pesticides, residues, adsorbed phosphorus, nitrogen, and other organic compounds, and pathogenic bacteria and viruses.

The problems caused by sediment deposition are varied. Sediments deposited in stream channels reduce flood-carrying capacity, resulting in more frequent overflows and greater floodwater damage to adjacent properties. The deposition of sediments in irrigation and drainage canals, in navigation channels and floodways, in reservoirs and harbors, on streets and highways, and in buildings not only creates a nuisance but inflicts a high public cost in maintenance removal or in reduced services.

Sedimentation is of vital concern in the conservation, development, and utilization of our soil and water resources. With a rapidly expanding population and an ever-increasing demand for food and products derived from soil and water, exploitation and apathy must rapidly be replaced by wise planning and circumspection if future generations are to maintain the standard of living prevalent today.

The physical analysis of erosion and sedimentation in this book rests on Newtonian mechanics applied to the motion of fluids and sediment particles. Chapter 2 outlines the physical properties of sediments and dimensional analysis. Chapter 3 encompasses the fundamental principles of fluid mechanics applied to sediment-laden flows. Chapter 4 deals with the motion of single particles in inviscid fluids, while Chapter 5 deals with the case of viscous fluids. The concept of turbulence and applications to sediment-laden flows are summarized in Chapter 6. Chapter 7 extends the analysis of beginning of motion of single particles to the three-dimensional case and presents applications to stable channel design. The complex topics of bedform configuration and resistance to flow are reviewed in Chapter 8, on both a conceptual and an empirical basis. The general topic of sediment transport is divided into three chapters: bedload in Chapter 9, suspended load in Chapter 10, and total load in Chapter 11. Sedimentation is covered in Chapter 12 with emphasis on reservoirs.

2

Physical properties and dimensional analysis

The processes of erosion, transport, and deposition of sediment particles introduced in Chapter 1 are related to the interaction between solid particles and the surrounding fluid. This chapter describes the physical properties of water and solid particles in terms of dimensions and units (Section 2.1), as well as the fundamental properties of water (Section 2.2) and of sediments (Section 2.3). The method of dimensional analysis is then applied to representative erosion and sedimentation problems (Section 2.4).

2.1   Dimensions and units

The physical properties of fluids and solids are usually expressed in terms of the following fundamental dimensions: mass (M), length (L), time (T), and temperature ($T°$). Some systems of units have replaced the unit of mass by a corresponding unit of force. The fundamental dimensions are measurable parameters that can be quantified in fundamental units.

In the Système International (SI), the basic *units* of mass, length, time, and temperature are the kilogram (kg), meter (m), second (s), and kelvin (K). Alternatively, the Celsius scale (°C) is commonly preferred. Accordingly, the freezing point of water is 0°C, and the boiling point is 100°C.

A newton (N) is defined as the force required to accelerate 1 kilogram at 1 meter per second squared. Given that the acceleration due to gravity at the earth's surface g is 9.81 m/s², the weight of a kilogram is obtained from Newton's second law: $F = $ mass $\times g = 1$ kg $\times 9.81$ m/s² $= 9.81$ N. The unit of work (or energy) is the joule (J), which equals the product of 1 newton times 1 meter. The unit of power is a watt (W), which is 1 joule per second. Prefixes are used in the SI system to indicate multiples or fractions of units by powers of 10:

$$
\left.
\begin{array}{l}
\mu \text{ (micro)} = 10^{-6} \\
\text{m (milli)} = 10^{-3} \\
\text{c (centi)} = 10^{-2}
\end{array}
\right\}
\left\{
\begin{array}{l}
\text{k (kilo)} = 10^{3} \\
\text{M (mega)} = 10^{6} \\
\text{G (giga)} = 10^{9}
\end{array}
\right.
$$

4

Table 2.1. *Geometric, kinematic, dynamic, and dimensionless variables*

Variable	Symbol	Fundamental dimensions	SI units
Geometric (L)			
Length	L, x, h, d_s	L	m
Area	A	L^2	m²
Volume	$\forall$	L^3	m³
Kinematic (L, T)			
Velocity	v_x, u, u_*	LT^{-1}	m/s
Acceleration	a, a_x, g	LT^{-2}	m/s²
Kinematic viscosity	ν	L^2T^{-1}	m²/s
Dynamic (M, L, T)			
Mass	m	M	1 kg
Force	$F = ma, mg$	MLT^{-2}	1 kg × m/s² = 1 newton
Pressure	$p = F/A$	$ML^{-1}T^{-2}$	1 N/m² = 1 pascal
Shear stress	$\tau_{xy}, \tau_0, \tau_c$	$ML^{-1}T^{-2}$	1 N/m² = 1 pascal
Work or energy	$E = F \times d$	ML^2T^{-2}	1 N × m = 1 joule
Power	$P = E/t$	ML^2T^{-3}	1 N × m/s = 1 watt
Mass density	ρ, ρ_s	ML^{-3}	kg/m³
Specific weight	$\gamma, \gamma_s = \rho_s g$	$ML^{-2}T^{-2}$	N/m³
Dynamic viscosity	$\mu = \rho\nu$	$ML^{-1}T^{-1}$	1 kg/m × s = 1 N × s/m² = 1 Pa
Dimensionless			
Slope	S_0, S_f	—	—
Specific gravity	$G = \gamma_s/\gamma$	—	—
Reynolds number	$\text{Re} = uh/\nu$	—	—
Grain shear Reynolds number	$\text{Re}_* = u_* d_s/\nu$	—	—
Froude number	$\text{Fr} = u/\sqrt{gh}$	—	—
Shields parameter	$\tau_* = \tau/(\gamma_s - \gamma)d_s$	—	—
Concentration	C	—	—

For example, 1 millimeter (mm) stands for 0.001 m and 1 megawatt (MW) equals 1 million watts.

In the English system of units, the time unit is a second; the fundamental units of length and mass are, respectively, the foot (ft), equal to 30.48 cm, and the slug, equal to 14.59 kg. The force required to accelerate a mass of 1 slug at 1 foot per second squared is a pound-force (lb) used throughout this text. The temperature in degrees Celsius $T°$ (°C) is converted to the temperature in degrees Fahrenheit $T°$ (°F) using $T°$ (°F) = 32.2°F + $1.8T°$ (°C).

Most physical variables can be described in terms of three fundamental dimensions (M, L, T). Variables are classified as geometric, kinematic, dynamic, and dimensionless, as shown in Table 2.1. Geometric variables

Table 2.2. *Conversion of units*

Unit	kg, m, s	N, Pa, W
1 acre	4046.87 m²	
1 atmosphere (atm)	101,325 kg/m × s²	101.3 kPa
1 Btu = 778 lb × ft	1,055 kg × m²/s²	1055 N × m
1 bar	100,000 kg/m × s²	100 kPa
1 day (d)	86,400 s	
1 dyne (dyn)	0.00001 kg × m/s²	1×10^{-5} N
1 dyn/cm²	0.1 kg/m × s²	0.1 Pa
1 fathom	1.8288 m	
1 foot (ft)	0.3048 m	
1 ft³/s	0.0283 m³/s	
1 U.S. gallon (gal)	0.0037854 m³	
1 mgd = 1 million gal/day = 1.55 ft³/s	0.04382 m³/s	
1 horsepower = 550 lb × ft/s	745.70 kg × m²/s³	745.7 W
1 inch (in.)	0.0254 m	
1 in. of mercury	3386.39 kg/m × s²	3386.39 Pa
1 in. of water	248.84 kg/m × s²	248.84 Pa
1 joule	1 kg m²/s²	1 N × m = 1 J
1 kip = 1000 lb	4448.22 kg × m/s²	4448.22 N
1 knot	0.5144 m/s	
1 liter (l)	0.001 m³	
1 mile (mi) = 5280 ft	1609.34 m	
1 micrometer (μm)	1×10^{-6} m	
1 nautical mile	1,852 m	
1 newton (N)	1 kg × m/s²	1 N
1 ounce (oz)	0.02835 kg	
1 fluid ounce	2.957×10^{-5} m³	
1 pascal (Pa)	1 kg/m × s²	1 N/m² = 1 Pa
1 Poise (P)	0.1 kg/m × s	0.1 Pa × s
1 pound-force (lb)	4.448 kg × m/s²	4.448 N
1 lb-ft	1.356 kg × m²/s²	1.356 N × m
1 psf (lb/ft²)	47.88 kg/m × s²	47.88 Pa
1 psi (lb/in.²)	6894.76 kg/m × s²	6894.76 Pa
1 pound-force per ft³	157.09 kg/m² s²	157.09 N/m³
1 pint (pt)	0.0004732 m³	
1 quart (qt)	0.00094635 m³	
1 slug	14.59 kg	
1 slug/ft³	515.4 kg/m³	
1 stokes (St) = 1 cm²/s	0.0001 m²/s	
1 metric ton	1000 kg	
1 short ton	907.18 kg	
1 long ton (U.K.)	1016.05 kg	
1 watt	1 kg m²/s³	1 W
1 yard	0.9144 m	
1 year	31,536,000 s	
1 degree Celsius (°C) = $(T\,°F - 32°)5/9$	1 K	
1 degree Fahrenheit (°F) = $32 + 1.8T\,°C$	0.555556 K	

involve length dimensions only and describe the geometry of a system through length, area, and volume. Kinematic variables describe the motion of fluid and solid particles and can be depicted by only two fundamental dimensions, namely L and T. Dynamic variables involve mass terms in the fundamental dimensions. Force, pressure, shear stress, work, energy, power, mass density, specific weight, and dynamic viscosity are common examples. Several conversion factors are listed in Table 2.2.

2.2   Physical properties of water

The principal properties of a nearly incompressible fluid like water are sketched in Figure 2.1.

Mass density of a fluid, ρ. The mass of fluid per unit volume is referred to as the mass density and is given the Greek symbol ρ. The mass density of water at 10°C is 1,000 kg/m^3 and varies slightly with temperature, as shown in Table 2.3. The maximum density of water is obtained at a temperature of 4°C (1 slug/ft^3 = 515.4 kg/m^3).

Specific weight of a fluid, γ. The gravitational force per unit volume of fluid, or simply the fluid weight per unit volume of fluid, defines the specific weight, described by the Greek symbol γ. At 10°C, water has a specific weight $\gamma = 9,810$ N/m^3, or 62.4 lb/ft^3 (1 lb/ft^3 = 157.09 N/m^3). Specific weight varies slightly with temperature, as indicated in Table 2.3. Mathematically, the specific weight γ equals the product of the mass density ρ times the gravitational acceleration $g = 32.2$ ft/s^2 = 9.81 m/s^2:

$$\gamma = \rho g \qquad (2.1)$$

Dynamic viscosity, μ. As a fluid is brought into deformation, the velocity of the fluid at any boundary equals the velocity of the boundary. The ensuing rate of fluid deformation causes a shear stress τ_{yx} proportional to the dynamic viscosity μ and the rate of deformation of the fluid, dv_x/dy:

$$\tau_{yx} = \mu \frac{dv_x}{dy} \qquad (2.2)$$

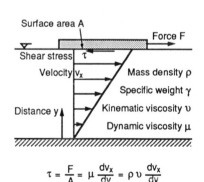

$$\tau = \frac{F}{A} = \mu \frac{dv_x}{dy} = \rho v \frac{dv_x}{dy}$$

Figure 2.1. Newtonian fluid properties

Table 2.3. *Approximate physical properties of clear water at atmospheric pressure*

Temp. (°C)	Density ρ (kg/m³)	Specific weight γ (N/m³)	Dynamic viscosity μ (N×s/m²)	Kinematic viscosity ν (m²/s)
0	1,000	9,810	1.79×10^{-3}	1.79×10^{-6}
5	1,000	9,810	1.51×10^{-3}	1.51×10^{-6}
10	1,000	9,810	1.31×10^{-3}	1.31×10^{-6}
15	999	9,800	1.14×10^{-3}	1.14×10^{-6}
20	998	9,790	1.00×10^{-3}	1.00×10^{-6}
25	997	9,781	8.91×10^{-4}	8.94×10^{-7}
30	996	9,771	7.97×10^{-4}	8.00×10^{-7}
35	994	9,751	7.20×10^{-4}	7.25×10^{-7}
40	992	9,732	6.53×10^{-4}	6.58×10^{-7}
50	988	9,693	5.47×10^{-4}	5.53×10^{-7}
60	983	9,643	4.66×10^{-4}	4.74×10^{-7}
70	978	9,594	4.04×10^{-4}	4.13×10^{-7}
80	972	9,535	3.54×10^{-4}	3.64×10^{-7}
90	965	9,467	3.15×10^{-4}	3.26×10^{-7}
100	958	9,398	2.82×10^{-4}	2.94×10^{-7}
(°F)	(slug/ft³)	(lb/ft³)	(lb×s/ft²)	(ft²/s)
40	1.94	62.43	3.23×10^{-5}	1.66×10^{-5}
50	1.94	62.40	2.73×10^{-5}	1.41×10^{-5}
60	1.94	62.37	2.36×10^{-5}	1.22×10^{-5}
70	1.94	62.30	2.05×10^{-5}	1.06×10^{-5}
80	1.93	62.22	1.80×10^{-5}	0.930×10^{-5}
100	1.93	62.00	1.42×10^{-5}	0.739×10^{-5}
120	1.92	61.72	1.17×10^{-5}	0.609×10^{-5}
140	1.91	61.38	0.981×10^{-5}	0.514×10^{-5}
160	1.90	61.00	0.838×10^{-5}	0.442×10^{-5}
180	1.88	60.58	0.726×10^{-5}	0.385×10^{-5}
200	1.87	60.12	0.637×10^{-5}	0.341×10^{-5}
212	1.86	59.83	0.593×10^{-5}	0.319×10^{-5}

The fundamental dimensions of the dynamic viscosity μ are M/LT, which is a dynamic variable. As indicated in Table 2.3, the dynamic viscosity of water decreases with temperature.

Fluids without yield stress for which the dynamic viscosity remains constant regardless of the rate of deformation are called Newtonian fluids. The dynamic viscosity of clear water at 20°C is 1 centipoise: 1 cP = 0.01 P = 0.001 N×s/m² = 0.001 Pa×s (1 lb×s/ft² = 47.88 N×s/m² = 47.88 Pa×s).

Kinematic viscosity, ν. When the dynamic viscosity of a fluid μ is divided by the mass density ρ of the same fluid, the mass terms cancel out. This results in the kinematic viscosity ν with dimensions L^2/T, which is also shown in Table 2.3 to decrease with temperature. The viscosity of clear water at 20°C is 1 centistoke $= 0.01$ cm^2/s $= 1 \times 10^{-6}$ m^2/s (1 ft^2/s $= 0.0929$ m^2/s). The change in the kinematic viscosity of water ν with temperature $T°$ in degrees Celsius can be roughly estimated from

$$\nu = \frac{\mu}{\rho} = [1.14 - 0.031(T° - 15) + 0.00068(T° - 15)^2] \times 10^{-6} \text{ m}^2/\text{s} \qquad (2.3)$$

It is important to remember that both the density and viscosity of water decrease with temperature.

2.3   Physical properties of sediment

The physical properties of sediment are classified into those of a single particle (Section 2.3.1), sediment mixture (Section 2.3.2), and sediment suspension (Section 2.3.3).

2.3.1   Single particle

The physical properties of a single solid particle of volume $\forall_s$ are sketched in Figure 2.2.

Mass density of solid particles, ρ_s. The mass density of a solid particle, ρ_s, describes the solid mass per unit volume. The mass density of quartz particles, 2,650 kg/m^3 (1 slug/ft^3 = 515.4 kg/m^3), does not vary significantly with temperature and is assumed constant in most calculations. It must be kept in mind, however, that heavy minerals like iron and copper have much larger values of mass density.

Specific weight of solid particles, γ_s. The particle specific weight, γ_s, corresponds to the solid weight per unit volume of solid. Typical values of γ_s are 26.0 kN/m^3 and 165.4 lb/ft^3 (1 lb/ft^3 = 157.09 N/m^3). The specific weight of a solid, γ_s, also equals the product of the mass density of a solid particle, ρ_s, times

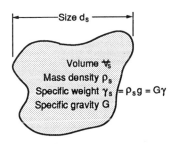

Figure 2.2. Physical properties of a single particle

Table 2.4. *Sediment grade scale*

Class name	Size range	
	mm	in.
Boulder		
Very large	4,096–2,048	160–80
Large	2,048–1,024	80–40
Medium	1,024–512	40–20
Small	512–256	20–10
Cobble		
Large	256–128	10–5
Small	128–64	5–2.5
Gravel		
Very coarse	64–32	2.5–1.3
Coarse	32–16	1.3–0.6
Medium	16–8	0.6–0.3
Fine	8–4	0.3–0.16
Very fine	4–2	0.16–0.08
Sand		
Very coarse	2.000–1.000	
Coarse	1.000–0.500	
Medium	0.500–0.250	
Fine	0.250–0.125	
Very fine	0.125–0.062	
Silt		
Coarse	0.062–0.031	
Medium	0.031–0.016	
Fine	0.016–0.008	
Very fine	0.008–0.004	
Clay		
Coarse	0.004–0.0020	
Medium	0.0020–0.0010	
Fine	0.0010–0.0005	
Very fine	0.0005–0.00024	

the gravitational acceleration g; thus,

$$\gamma_s = \rho_s g \tag{2.4}$$

Submerged specific weight of a particle, γ_s'. Owing to Archimedes' principle, the specific weight of a solid particle, γ_s, submerged in a fluid of specific weight γ equals the difference between the two specific weights; thus,

$$\gamma_s' = \gamma_s - \gamma \tag{2.5}$$

Specific gravity, G. The ratio of the specific weight of a solid particle to the specific weight of a fluid at a standard reference temperature defines the specific gravity G. With common reference to water at 4°C, the specific gravity of quartz particles is

$$G = \frac{\gamma_s}{\gamma} = \frac{\rho_s}{\rho} = 2.65 \qquad (2.6)$$

The specific gravity is a dimensionless ratio of specific weights, and thus its value remains independent of the system of units.

Sediment size, d_s. The most important physical property of a sediment particle is its size. Table 2.4 shows the grade scale commonly used in sedimentation. Note that the size scales are arranged in geometric series with a ratio of 2 (1 in. = 25.4 mm).

The size of particles can be determined in a number of ways. The nominal diameter refers to the diameter of a sphere of same volume as the particle, usually measured by the displaced volume of a submerged particle. The sieve diameter is the minimum length of the square sieve opening through which a particle will fall. The fall diameter is the diameter of an equivalent sphere of specific gravity $G = 2.65$ having the same terminal settling velocity in water at 24°C.

Particle shape factor, Co. The shape of sediment particles can be described by measures of the longest axis l_a, the intermediate l_b, and shortest axis l_c. The Corey shape factor $Co = l_c/\sqrt{l_a l_b}$ is always less than unity, and values of 0.7 are typical for naturally worn particles.

2.3.2   Sediment mixture

The properties of a sediment mixture are sketched in Figure 2.3.

Particle size distribution. The particle size distribution in Figure 2.4 shows the percentage by weight of material finer than a given sediment size. The sediment size d_{50} for which 50% by weight of the material is finer is called the *median grain size*. Likewise d_{90} and d_{10} are values of grain size for which 90% and 10% of the material weight is finer, respectively.

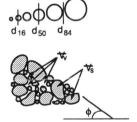

Figure 2.3. Properties of a sediment mixture

Sieve analysis. Sieving is considered a semidirect method of particle size measurement

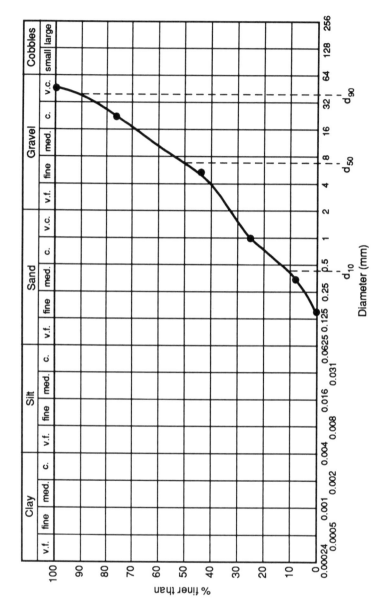

Figure 2.4.   Particle size distribution (v.f., very fine; c., coarse; v.c., very coarse)

because it does not divide the particles precisely. There are at least four reasons for this: (1) irregularities in shape – particles larger or smaller (larger for cylinders and smaller for disks) than the spherical equivalent diameter may pass through the sieve openings; (2) inaccuracies in the size and shape of the sieve openings; (3) the failure of some particles to pass a given sieve opening owing to the finite duration of the sieving operation; and (4) the clinging of small particles to large ones, thereby changing the percentage of material reaching the sieves with smaller openings.

A wet-sieve method keeps the sieve screen and sand completely submerged. The equipment may consist of six or more 10-cm ceramic dishes, a set of 3-in. (7.5-cm) sieves, and a thin glass tube. All sieves are washed with a wetting solution (detergent) and then raised gently with distilled water so that a membrane of water remains across all openings. The first, or largest, sieve is immersed in a ceramic dish with distilled water to a depth of about ¼ in. (½ cm) above the screen. The sediment is washed onto the wet sieve and agitated somewhat vigorously in several directions until all particles smaller than the sieve openings have a chance to fall through the sieve. Material passing through the sieve with its wash water is then poured onto a sieve of the next smaller size. Particles retained on each sieve and those passing the 0.062-mm sieve are transferred to containers that are suitable for drying the material and for obtaining the net weight of each fraction.

The dry-sieve method is less laborious than the wet-sieve method because a mechanical shaker can be used with a nest of sieves for simultaneous separation of all sizes of interest. It requires only that the dry sand be poured over the coarsest sieve and the nest of sieves shaken for 10 min on a shaker having both lateral and vertical movements.

Gradation coefficients, σ_g and Gr. The gradation of the sediment mixture can be described by the standard deviation

$$\sigma_g = \left(\frac{d_{84}}{d_{16}}\right)^{1/2} \tag{2.7a}$$

or by the gradation coefficient

$$Gr = \frac{1}{2}\left(\frac{d_{84}}{d_{50}} + \frac{d_{50}}{d_{16}}\right) \tag{2.7b}$$

Angle of repose, ϕ. The angle of repose of submerged loose material is the sideslope angle, with respect to the horizontal, of a cone of material under incipient sliding conditions. The angle of repose of granular material ranges between 30° and 42°. Specific measurements of repose are given in Chapter 7.

Porosity, p_0. The porosity p_0 is a measure of the volume of void $\forall_v$ per total volume $\forall_t = \forall_v + \forall_s$. The volume of solid particles $\forall_s = (1 - p_0)\forall_t$; thus,

$$p_0 = \frac{\forall_v}{\forall_t} = \frac{e}{1+e} \tag{2.8}$$

Void ratio, e. The void ratio e is a measure of the volume of void $\forall_v$ per volume of solid $\forall_s$, or

$$e = \frac{\forall_v}{\forall_s} = \frac{p_0}{1-p_0} \tag{2.9}$$

Dry specific weight of a mixture, γ_{md}. The dry specific weight of a mixture is the weight of solid per unit total volume, including the volume of solids and voids. The dry specific weight of a mixture γ_{md} is a function of porosity p_0 as

$$\gamma_{md} = \gamma_s(1 - p_0) = \gamma G(1 - p_0) \tag{2.10}$$

Dry specific mass of a mixture, ρ_{md}. The dry specific mass of a mixture is the mass of solid per unit total volume. The dry specific mass of a mixture can be defined as a function of porosity p_0 as

$$\rho_{md} = \rho_s(1 - p_0) = \frac{\gamma_{md}}{g} \tag{2.11}$$

2.3.3   Sediment suspension

The properties of a sediment suspension are sketched in Figure 2.5, with the volume of void $\forall_v$ equal to the volume of water $\forall_w$.

Specific weight of a mixture, γ_m. The specific weight of a submerged mixture is the total weight of solid and water in the voids per unit total volume. The specific weight of a mixture γ_m is a function of porosity

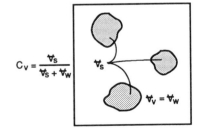

Figure 2.5. Properties of a suspension

p_0 as

$$\gamma_m = \frac{\gamma_s \mathbf{V}_s + \gamma \mathbf{V}_v}{\mathbf{V}_s + \mathbf{V}_v} = \gamma_s(1-p_0) + \gamma p_0 = \gamma(G(1-p_0)+p_0) \qquad (2.12)$$

Specific mass of a mixture, ρ_m. The specific mass of a submerged mixture is the total mass of solid and water in the voids per unit total volume. The specific mass of a mixture ρ_m can be defined as a function of porosity p_0 as

$$\rho_m = \rho_s(1-p_0) + \rho p_0 = \frac{\gamma_m}{g} \qquad (2.13)$$

Volumetric sediment concentration, C_v. The volumetric sediment concentration C_v is defined as the volume of solids $\mathbf{V}_s$ over the total volume $\mathbf{V}_t$, or

$$C_v = \frac{\mathbf{V}_s}{\mathbf{V}_t} = \frac{\mathbf{V}_s}{\mathbf{V}_s + \mathbf{V}_v} \qquad (2.14a)$$

By analogy with the definition of porosity, when the voids are completely filled with water, $\mathbf{V}_v = \mathbf{V}_w$,

$$C_v = \frac{\mathbf{V}_s}{\mathbf{V}_s + \mathbf{V}_w} = 1 - p_0 \qquad (2.14b)$$

Conversions to concentration by weight, C_w, C_{ppm}, and $C_{mg/l}$, are presented in Section 10.1.

Dynamic viscosity of a Newtonian mixture, μ_m. The dynamic viscosity of a Newtonian mixture, μ_m, is the ratio of applied shear stress to the rate of deformation of a Newtonian suspension. A. Einstein suggested the following function of volumetric sediment concentration C_v:

$$\mu_m = \mu(1 + 2.5C_v) \qquad (2.15)$$

This relationship is approximate, even at low volumetric concentrations ($C_v < 0.05$).

Kinematic viscosity of a Newtonian mixture, $\nu_m = \mu_m/\rho_m$. The kinematic viscosity of a Newtonian mixture, ν_m, is obtained by dividing the dynamic viscosity of a Newtonian mixture, μ_m, by the mass density of the mixture, ρ_m.

2.4   Dimensional analysis

The quantification of geometric, kinematic, and dynamic variables involves the use of a consistent system of units. While the numerical value of a dimensional variable depends on the system of units, dimensionless variables remain invariant in all systems. Dimensional variables can be combined to give dimensionless parameters through dimensional analysis.

Dimensional analysis is a method by which we deduce information about a phenomenon with the single premise that it can be described by a dimensionally correct equation among certain variables. Dimensional analysis meets the double objective of (1) reducing the number of variables for subsequent analysis and (2) providing dimensionless parameters whose numerical values are independent of any system of units. Dimensional reasoning by itself will lead neither to a complete solution to a physical investigation nor to a clear understanding of an inner mechanism. Dimensional analysis is, however, a useful tool for dealing with the mathematics of the dimensions of quantities. The following presentation combines the methods of Rayleigh, Buckingham, and Hunsaker and Rightmire.

Buckingham's Π theorem allows us to rearrange n variables in which there are j fundamental dimensions into $n-j$ dimensionless parameters designated by the Greek letter Π. Let the dependent variable Z_1 be related by any functional relationship $\mathcal{F}$ to the independent variables $Z_2, ..., Z_n$ such that

$$Z_1 = \mathcal{F}(Z_2, ..., Z_n) \qquad (2.16a)$$

with j fundamental dimensions involved, for example, $j = 3$, when their n parameters combine mass M, length L, and time T. The relation can be reduced to a function of $n-j$ dimensionless Π parameters in the form

$$\Pi_1 = \mathcal{F}(\Pi_2, \Pi_3, ..., \Pi_{n-j}) \qquad (2.16b)$$

The method of determining the Π parameters is to select j repeating variables among the Z variables such that (1) all j fundamental dimensions can be found in the set of repeating variables and (2) all repeating variables have different fundamental dimensions. For instance, one cannot select both the width and the length as repeating variables, because both parameters have the same fundamental dimension L.

The steps in a dimensional analysis can be summarized as follows:

1. Select the dependent variable Z_1 as a function of the independent variables $Z_2, ..., Z_n$ in the functional relationship

   $Z_1 = \mathcal{F}(Z_2, Z_3, ..., Z_n)$.

2. Write the variables in terms of fundamental dimensions and select the repeating variables. These variables must contain the j dimensions of the problem, and the dependent quantity should not be selected as a repeating variable.

3. Write the Π parameters in terms of fundamental dimensions and substitute the corresponding functions of repeating variables.
4. Write the functional relation

$$\mathcal{F}(\Pi_1, \Pi_2, ..., \Pi_{n-j}) \quad \text{or} \quad \Pi_1 = \mathcal{F}(\Pi_2, ..., \Pi_{n-j})$$

and recombine if desired to alter the form of the dimensionless parameters Π, keeping the same number of independent parameters.

The method of dimensional analysis is illustrated in the following example of the drag force exerted on a sphere in relative fluid motion (Example 2.1). An analysis of soil erosion by overland flow based on the method of dimensional analysis (Example 2.2) follows.

Example 2.1   **Application to drag force on a sphere.** Consider the drag force F_D exerted on a sphere in motion through a homogeneous fluid-sediment mixture (Fig. E2.1.1).

Step 1. The dependent variable F_D is a function of four independent variables, with a total of $n = 5$ variables:

$$F_D = \mathcal{F}(u_\infty, d_s, \rho_m, \mu_m) \tag{E2.1.1}$$

in which $\mathcal{F}$ represents an unspecified function of the relative velocity u_∞, the spherical particle diameter d_s, the mass density of the fluid mixture ρ_m, and the dynamic viscosity of the mixture μ_m.

Step 2. After d_s, u_∞, and ρ_m are selected as repeating variables, the three fundamental dimensions ($j = 3$) are rewritten in terms of repeating variables:

$$\left. \begin{array}{l} d_s = L \\ u_\infty = L/T \\ \rho_m = M/L^3 \end{array} \right\} \quad \text{thus} \quad \left\{ \begin{array}{l} L = d_s \\ T = d_s/u_\infty \\ M = \rho_m d_s^3 \end{array} \right.$$

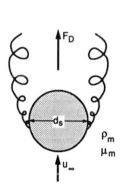

Step 3. After the relationships for M, L, T are substituted into the nonrepeating variables, two Π terms ($n - j = 5 - 3 = 2$) are obtained, respectively, from F_D and μ_m:

$$\Pi_1: \quad F_D = \frac{ML}{T^2} = \frac{\rho_m d_s^3 d_s u_\infty^2}{d_s^2}$$

or

Figure E2.1.1. Drag force on a moving sphere

$$\Pi_1 = \frac{F_D}{\rho_m u_\infty^2 d_s^2} = \frac{\pi}{8} C_D \tag{E2.1.2}$$

Π_2: $\mu_m = \dfrac{M}{LT} = \dfrac{\rho_m d_s^3}{d_s d_s} u_\infty$

or

$$\Pi_2 = \frac{\rho_m u_\infty d_s}{\mu_m} = \mathrm{Re}_p$$

Step 4. $\Pi_1 = \mathfrak{F}(\Pi_2)$, or $C_D = \mathfrak{F}(\mathrm{Re}_p)$, results in

$$F_D = \mathfrak{F}(\mathrm{Re}_p)\frac{\pi d_s^2}{4}\frac{\rho_m u_\infty^2}{2} \qquad\qquad (E2.1.3) \; \blacklozenge$$

The advantage of using dimensional analysis in this case is that the number of parameters is reduced from five in Equation (E2.1.1) to two dimensionless parameters, $\Pi_1 = C_D$ and $\Pi_2 = \mathrm{Re}_p$. Note that the particle Reynolds number describes the ratio of kinetic to viscous forces applied on a moving particle. The method, however, fails to indicate what kind of relationship may exist between C_D and Re_p. For further analysis, the scientist must carry out experiments and collect laboratory or field measurements of C_D versus Re_p. Measurements of drag coefficients around spheres versus the particle Reynolds number $\mathrm{Re}_p = u_\infty d_s/\nu_m$ are shown in Figure E2.1.2. Note that a similar plot for natural sand particles is also shown in Figure 5.2. The flow around a particle is said to be laminar when $\mathrm{Re}_p < 1$ and turbulent at large particle Reynolds numbers.

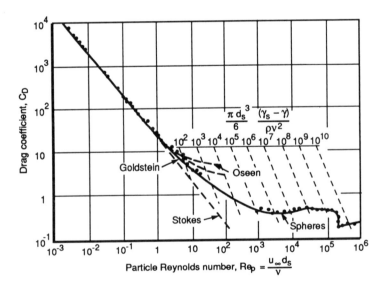

Figure E2.1.2. Drag coefficient for spheres

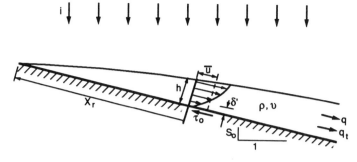

Figure E2.2.1. Sheet erosion

Example 2.2 Analysis of soil erosion by overland flow. Consider the problem of sheet erosion induced by rainfall on a bare soil surface (Fig. E2.2.1). The method of dimensional analysis is first used to reduce the number of variables and define dimensionless parameters.

Step 1. The rate of sediment transport by sheet erosion q_t is written as a function of the geometric, fluid flow, and soil variables,

$$q_t = \mathfrak{F}\left(S_0, q, i, X_r, \rho, \nu, \frac{\tau_c}{\tau_0}\right) \tag{E2.2.1}$$

in which q_t is the rate of mass transport per unit width, q is the unit discharge, i is the rainfall intensity, X_r is the length of runoff, ρ is the mass density of the fluid, ν is the kinematic viscosity of the fluid, τ_c and τ_0 are, respectively, the critical and applied boundary shear stress, and S_0 is the bed surface slope.

Besides the two dimensionless variables in Equation E2.2.1 ($\Pi_5 = \tau_c/\tau_0$, $\Pi_2 = S_0$), the other six variables ($n = 6$) are functions of three fundamental dimensions (M, L, T; thus, $j = 3$) and can be transformed into three ($n - j = 3$) dimensionless parameters. Each variable is written in terms of the fundamental dimensions M, L, and T as follows:

$$q_t = M/LT$$
$$q = L^2/T$$
$$i = L/T$$
$$X_r = L$$
$$\nu = L^2/T$$
$$\rho = M/L^3$$

Step 2. The fundamental dimensions can be written in terms of the repeated variables X_r, ν, and ρ:

$$
\left.\begin{array}{l}
X_r = L \\
\nu = L^2/T \\
\rho = M/L^3
\end{array}\right\} \text{ thus } \left\{\begin{array}{l}
M = \rho X_r^3 \\
L = X_r \\
T = X_r^2/\nu
\end{array}\right.
$$

Step 3. The three Π parameters are directly obtained by substituting the fundamental dimensions into the relationships for q_t, q, and i, respectively:

$$
\Pi_1 = \frac{q_t}{M}LT = \frac{q_t X_r X_r^2}{\rho X_r^3 \nu} = \frac{q_t}{\rho \nu}
$$

$$
\Pi_3 = \frac{qT}{L^2} = \frac{qX_r^2}{X_r^2 \nu} = \frac{q}{\nu} = \text{Re}
$$

$$
\Pi_4 = \frac{iT}{L} = \frac{iX_r^2}{X_r \nu} = \frac{iX_r}{\nu}
$$

Step 4. The five dimensionless parameters can thus be written

$$
\frac{q_t}{\rho \nu} = \mathcal{F}\left(S_0, \frac{q}{\nu}, \frac{iX_r}{\nu}, \frac{\tau_c}{\tau_0}\right) \tag{E2.2.2} \blacklozenge
$$

The results from this dimensional analysis indicate that the dimensionless sediment transport parameter is a function of the soil surface slope, the Reynolds number, a dimensionless rainfall parameter, and the soil characteristics.

Further progress can be achieved only through physical understanding of the erosion processes and laboratory or field experiments. For instance, the rate of sediment transport in sheet flow is assumed to be proportional to the product of the powers of the dimensionless parameters

$$
\left(\frac{q_t}{\rho \nu}\right) = e_1 S_0^{e_2}\left(\frac{q}{\nu}\right)^{e_3}\left(\frac{iX_r}{\nu}\right)^{e_4}\left(1 - \frac{\tau_c}{\tau_0}\right)^{e_5} \tag{E2.2.3}
$$

in which e_1, e_2, e_3, e_4, and e_5 are coefficients to be determined from laboratory or field investigations.

The first three factors (S_0, q, i) of Equation (E2.2.3) represent the potential erosion or sediment transport capacity of overland flow. The sediment transport capacity is reduced by the last factor reflecting the soil resistance to erosion. When τ_c remains small compared with τ_0, at constant X_r and ν, Equation (E2.2.3) can be rearranged in the following form:

$$
q_t = \bar{e}_1 S_0^{e_2} q^{e_3} i^{e_4} \tag{E2.2.4}
$$

This sediment transport capacity relationship indicates that, for laminar sheet flow, the transport capacity depends primarily on the geometry of the upland area as described by the surface slope S_0, runoff discharge q, and rainfall intensity i. In the case of turbulent flow, this equation can be further reduced since rainfall impact is negligible when the flow depth is larger than three times the raindrop diameter; thus, $e_4 = 0$. At a given field site (constant slope S_0), Equation (E2.2.4) further reduces to

$$q_t \sim q^{e_3} \tag{E2.2.5}$$

This relationship, known as the sediment-rating curve, commonly describes the sediment transport capacity ($\tau_0 \gg \tau_c$) in alluvial channels. From field observations to be discussed in Chapter 11 (Section 11.2.1), the value of the exponent e_3 is typically of the order of 2 (e.g., Fig. 11.3).

Problem 2.1

Determine the mass density, specific weight, dynamic viscosity, and kinematic viscosity of clear water at 20°C (a) in SI units and (b) in the English system of units.

Answer

(a) $\rho = 998$ kg/m^3, $\gamma = 9790$ N/m^3, $\mu = 1.0 \times 10^{-3}$ N$\times$s/m^2, $\nu = 1 \times 10^{-6}$ m^2/s;
(b) $\rho = 1.94$ slug/ft^3, $\gamma = 62.3$ lb/ft^3, $\mu = 2.1 \times 10^{-5}$ lb$\times$s/m^2, $\nu = 1.1 \times 10^{-5}$ ft^2/s.

*Problem 2.2

Determine the sediment size, mass density, specific weight, and submerged specific weight of small quartz cobbles (a) in SI units and (b) in the English system of units.

**Problem 2.3

The volumetric sediment concentration of a sample is $C_v = 0.05$. Determine the corresponding porosity p_0; void ratio e; specific weight γ_m; specific mass ρ_m; dry specific weight γ_{md}; and dry specific mass ρ_{md}.

Answer: $p_0 = 0.95$; $e = 19$; $\gamma_m = 10.6$ kN/m^3; $\rho_m = 1082$ kg/m^3; $\gamma_{md} = 1.29$ kN/m^3; $\rho_{md} = 132$ kg/m^3.

**Problem 2.4

A 50-g bed-sediment sample is analyzed for particle size distribution.

Size fraction (mm)	Weight (mg)
$d_s < 0.15$	900
$0.15 < d_s < 0.21$	2,900
$0.21 < d_s < 0.30$	16,000
$0.30 < d_s < 0.42$	20,100
$0.42 < d_s < 0.60$	8,900
$0.60 < d_s$	1,200

(a) Plot the sediment size distribution;
(b) determine d_{16}, d_{35}, d_{50}, d_{65}, and d_{84}; and
(c) calculate the gradation coefficients σ_g and Gr.

**Problem 2.5

Consider energy losses in a straight open channel. The energy gradient $\Delta H_L / X_c$ in a smooth channel with turbulent flow depends on the mean flow velocity V, the flow depth h, the gravitational acceleration g, the mass density ρ, and the dynamic viscosity μ. Determine the general form of the energy gradient equation from dimensional analysis.

Answer

$$\frac{\Delta H_L}{X_c} = \mathcal{F}\left(\text{Reynolds number Re} = \frac{\rho V h}{\mu}; \text{ Froude number Fr} = \frac{V}{\sqrt{gh}}\right)$$

*Problem 2.6

Consider a near-bed turbulent velocity profile. The time-averaged velocity v at a distance y from the bed depends on the bed-material size d_s, the flow depth h, the dynamic viscosity of the fluid μ, the mass density ρ, and the boundary shear stress τ_0. Use the method of dimensional analysis to obtain a complete set of dimensionless parameters. [*Hint:* Select h, ρ, and τ_0 as repeating variables, and the problem reduces to a kinematic problem after defining the shear velocity $u_* = \sqrt{\tau_0/\rho}$ and $\nu = \mu/\rho$.]

Answer

$$\mathscr{F}\left(v\sqrt{\frac{\rho}{\tau_0}}, \frac{\rho d_s}{\mu}\sqrt{\frac{\tau_0}{\rho}}, \frac{y}{h}, \frac{d_s}{h}\right) = 0$$

or

$$\mathscr{F}\left(\frac{v}{u_*}, \frac{y}{h}, \text{ grain shear Reynolds number } \frac{u_* d_s}{\nu}, \text{ and } \frac{u_* y}{\nu}, \text{ or } \frac{d_s}{h}\right) = 0$$

3

Mechanics of sediment-laden flows

This chapter summarizes some fundamental principles in fluid mechanics applied to sediment-laden flows. The major topics reviewed include kinematics of flow (Section 3.1), continuity (Section 3.2), the equations of motion (Section 3.3), the Euler and Bernoulli equations (Sections 3.4 and 3.5), momentum equations (Section 3.6), and the energy equation expressing the rate of work done (Section 3.7). Seven examples illustrate various applications of these fundamental concepts.

3.1   Kinematics of flow

The kinematics of flow describes the motion in terms of velocity and acceleration of fluids under deformation. The three most common orthogonal coordinate systems are Cartesian (x, y, z), cylindrical (r, θ, z), and spherical (r, θ, φ), as shown in Figure 3.1.

The rate of change in the position of a fluid element is a measure of its velocity. Velocity is defined as the ratio between the displacement ds and

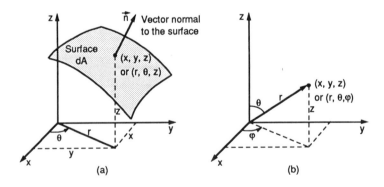

Figure 3.1. (a) Cartesian and cylindrical coordinates; (b) spherical coordinates

24

the corresponding increment of time dt. Velocity is a vector quantity $\vec{v}$ that varies in both space (x, y, z) and time (t). Its magnitude v at a given time equals the square root of the sum of squares of its orthogonal components, $v = \sqrt{v_x^2 + v_y^2 + v_z^2}$.

A line tangent to the velocity vector at every point at a given instant is known as a streamline. The path line of a fluid element is the locus of the element through time. A streak line is defined as the line connecting all fluid elements that have passed successively at a given point in space.

The differential velocity components over an infinitesimal distance ds (dx, dy, dz) and time increment dt at a point (x, y, z) are

$$dv_x = \frac{\partial v_x}{\partial t} dt + \frac{\partial v_x}{\partial x} dx + \frac{\partial v_x}{\partial y} dy + \frac{\partial v_x}{\partial z} dz \tag{3.1a}$$

$$dv_y = \frac{\partial v_y}{\partial t} dt + \frac{\partial v_y}{\partial x} dx + \frac{\partial v_y}{\partial y} dy + \frac{\partial v_y}{\partial z} dz \tag{3.1b}$$

$$dv_z = \frac{\partial v_z}{\partial t} dt + \frac{\partial v_z}{\partial x} dx + \frac{\partial v_z}{\partial y} dy + \frac{\partial v_z}{\partial z} dz \tag{3.1c}$$

$$\underbrace{}_{\text{local}} \quad \underbrace{}_{\text{convective}}$$

One may now consider translation, linear deformation, angular deformation, and rotation of a fluid element, as represented in Figure 3.2.

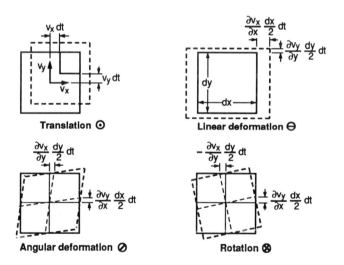

Figure 3.2. Translation, linear deformation, angular deformation, and rotation of a fluid element

Translation describes the displacement of the center of the fluid element. The rate of linear deformation is indicated by the quantities Θ_x, Θ_y, and Θ_z defined from the velocity components (v_x, v_y, v_z) as

$$\Theta_x = \frac{\partial v_x}{\partial x}; \qquad \Theta_y = \frac{\partial v_y}{\partial y}; \qquad \Theta_z = \frac{\partial v_z}{\partial z} \tag{3.2}$$

The velocity gradients in transverse directions (e.g., $\partial v_x/\partial y$) represent the rate of angular deformation of the element. The rates of angular deformation $\oslash$ in the respective planes are

$$\oslash_x = \frac{\partial v_z}{\partial y} + \frac{\partial v_y}{\partial z}; \qquad \oslash_y = \frac{\partial v_x}{\partial z} + \frac{\partial v_z}{\partial x}; \qquad \oslash_z = \frac{\partial v_y}{\partial x} + \frac{\partial v_x}{\partial y} \tag{3.3}$$

The mean rates of rotation $\otimes$ in their respective planes are defined as

$$\otimes_x = \left(\frac{\partial v_z}{\partial y} - \frac{\partial v_y}{\partial z} \right); \qquad \otimes_y = \left(\frac{\partial v_x}{\partial z} - \frac{\partial v_z}{\partial x} \right); \qquad \otimes_z = \left(\frac{\partial v_y}{\partial x} - \frac{\partial v_x}{\partial y} \right) \tag{3.4}$$

The components $\otimes_x$, $\otimes_y$, and $\otimes_z$ of the vorticity vector $\bar{\omega}$ equal the limiting value of the circulation Γ per unit area about their respective normal planes. Hence,

$$\Gamma = \int_A \left(\otimes_x \frac{\partial x}{\partial n} + \otimes_y \frac{\partial y}{\partial n} + \otimes_z \frac{\partial z}{\partial n} \right) dA \tag{3.5}$$

The properties of the vorticity vector $\bar{\omega}$ and the velocity vector $\bar{v}$ for homogeneous incompressible fluids are strikingly similar:

$$\text{div } \bar{v} = \frac{\partial v_x}{\partial x} + \frac{\partial v_y}{\partial y} + \frac{\partial v_z}{\partial z} = 0 \tag{3.6a}$$

$$\text{div } \bar{\omega} = \frac{\partial \otimes_x}{\partial x} + \frac{\partial \otimes_y}{\partial y} + \frac{\partial \otimes_z}{\partial z} = 0 \tag{3.6b}$$

Likewise, streamlines and vortex lines are respectively defined as

$$\frac{dx}{v_x} = \frac{dy}{v_y} = \frac{dz}{v_z} \tag{3.7a}$$

$$\frac{dx}{\otimes_x} = \frac{dy}{\otimes_y} = \frac{dz}{\otimes_z} \tag{3.7b}$$

Since the circulation around any cross section of a vortex tube is constant at a given instant, the vorticity is inversely proportional to the cross-sectional area of the tube and must end at a boundary or on itself (like a smoke ring).

The differential velocity components can be written as a function of local, linear, angular, and rotational acceleration terms:

$$dv_x = \frac{\partial v_x}{\partial t} dt + \Theta_x\, dx + \frac{\oslash_z}{2} dy + \frac{\oslash_y}{2} dz + \frac{1}{2}(\otimes_y\, dz - \otimes_z\, dy) \tag{3.8a}$$

$$dv_y = \frac{\partial v_y}{\partial t} dt + \Theta_y\, dy + \frac{\oslash_x}{2} dz + \frac{\oslash_z}{2} dx + \frac{1}{2}(\otimes_z\, dx - \otimes_x\, dz) \tag{3.8b}$$

$$dv_z = \frac{\partial v_z}{\partial t} dt + \Theta_z\, dz + \frac{\oslash_y}{2} dx + \frac{\oslash_x}{2} dy + \frac{1}{2}(\otimes_x\, dy - \otimes_y\, dx) \tag{3.8c}$$

$$\underbrace{\qquad}_{\text{local}} \quad \underbrace{\qquad}_{\text{linear}} \quad \underbrace{\qquad\qquad}_{\text{angular}} \quad \underbrace{\qquad\qquad}_{\text{rotational}}$$

The Cartesian acceleration components are obtained directly after rearranging the terms in the velocity equations [for example, one obtains Eq. (3.9a) from Eq. (3.8a) by adding and subtracting the terms $v_y\, \partial v_y/\partial x$ and $v_z\, \partial v_z/\partial x$],

$$a_x = \frac{dv_x}{dt} = \frac{\partial v_x}{\partial t} + v_x \frac{\partial v_x}{\partial x} + v_y \frac{\partial v_x}{\partial y} + v_z \frac{\partial v_x}{\partial z} = \frac{\partial v_x}{\partial t} + v_z \otimes_y - v_y \otimes_z + \frac{\partial (v^2/2)}{\partial x} \tag{3.9a}$$

$$a_y = \frac{dv_y}{dt} = \frac{\partial v_y}{\partial t} + v_x \frac{\partial v_y}{\partial x} + v_y \frac{\partial v_y}{\partial y} + v_z \frac{\partial v_y}{\partial z} = \frac{\partial v_y}{\partial t} + v_x \otimes_z - v_z \otimes_x + \frac{\partial (v^2/2)}{\partial y} \tag{3.9b}$$

$$a_z = \frac{dv_z}{dt} = \frac{\partial v_z}{\partial t} + v_x \frac{\partial v_z}{\partial x} + v_y \frac{\partial v_z}{\partial y} + v_z \frac{\partial v_z}{\partial z} = \frac{\partial v_z}{\partial t} + v_y \otimes_x - v_x \otimes_y + \frac{\partial (v^2/2)}{\partial z} \tag{3.9c}$$

$$\underbrace{\quad}_{\text{local}} \quad \underbrace{\qquad\qquad}_{\text{convective}} \quad \underbrace{\quad}_{\substack{\text{local}}} \quad \underbrace{\qquad}_{\substack{\text{convective} \\ \text{rotational}}} \quad \underbrace{\qquad}_{\substack{\text{convective} \\ \text{irrotational}}}$$

in which

$$v^2 = v_x^2 + v_y^2 + v_z^2$$

Equations (3.9) show that the total acceleration can be separated into local and convective acceleration terms, while the convective acceleration can itself be divided into rotational and irrotational components.

3.2 Equation of continuity

The equation of continuity is based on the law of conservation of mass, which states that mass cannot be created or destroyed. The continuity equation can be written in either differential or integral form.

3.2.1 Differential continuity equation

In differential form, consider the infinitesimal control volume in Figure 3.3 filled with fluid and sediment. The difference between the mass fluxes

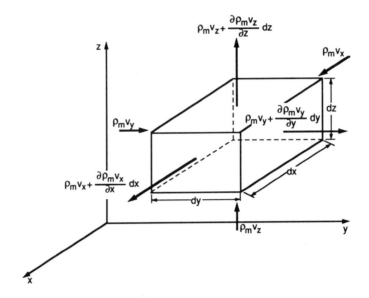

Figure 3.3. Infinitesimal element of fluid

entering and leaving the differential control volume equals the rate of increase of internal mass. The assumption of a continuous fluid medium yields the following differential relationships:

Cartesian coordinates (x, y, z)

$$\frac{\partial \rho_m}{\partial t} + \frac{\partial}{\partial x}(\rho_m v_x) + \frac{\partial}{\partial y}(\rho_m v_y) + \frac{\partial}{\partial z}(\rho_m v_z) = 0 \tag{3.10a}$$

Cylindrical coordinates (r, θ, z)

$$\frac{\partial \rho_m}{\partial t} + \frac{1}{r}\frac{\partial}{\partial r}(\rho_m r v_r) + \frac{1}{r}\frac{\partial}{\partial \theta}(\rho_m v_\theta) + \frac{\partial}{\partial z}(\rho_m v_z) = 0 \tag{3.10b}$$

Spherical coordinates (r, θ, φ)

$$\frac{\partial \rho_m}{\partial t} + \frac{1}{r^2}\frac{\partial}{\partial r}(\rho_m r^2 v_r) + \frac{1}{r \sin\theta}\frac{\partial}{\partial \theta}(\rho_m v_\theta \sin\theta)$$

$$+ \frac{1}{r \sin\theta}\frac{\partial}{\partial \varphi}(\rho_m v_\varphi) = 0 \tag{3.10c}$$

The conservation of solid mass is defined in Example 3.1. The continuity equations for solids are identical with Equations (3.10a–c) after replacing ρ_m with C_v. For homogeneous incompressible suspensions without settling, the mass density is independent of space and time ($\rho_s, \rho, \rho_m =$

constant); consequently, $\partial \rho_m / \partial t = 0$ and the divergence of the velocity vector in Cartesian coordinates must be zero:

$$\frac{\partial v_x}{\partial x} + \frac{\partial v_y}{\partial y} + \frac{\partial v_z}{\partial z} = 0 \qquad\qquad (3.10d) \;\blacklozenge$$

When one is dealing with open-channel flows with low sediment concentrations, compressibility effects are negligible and Equation (3.10d) is applicable.

Example 3.1 Derivation of the differential sediment continuity equation. Consider a sediment point source at a rate $\dot{m} = M/T$ added to the sediment suspension in a cubic element $dx\,dy\,dz$. Derive the governing sediment continuity equation given the volumetric sediment concentration C_v. The total mass of sediment inside the control volume is $dm = \rho_s C_v \,dx\,dy\,dz$. The mass change per unit time including the source is $dm/dt = \dot{m} - (d/dt)(\rho_s C_v)\,dx\,dy\,dz = 0$.

The mass flux entering the control volume by advection in the x direction is $\rho_s C_v v_x \,dy\,dz$, and the mass flux leaving the control volume in the x direction is $[\rho_s C_v v_x + (\partial/\partial x)(\rho_s C_v v_x)\,dx]\,dy\,dz$. With similar considerations in the y and z directions, the rate of mass change inside the control volume equals the net mass flux entering from all three directions; thus,

$$\frac{dm}{dt} = \dot{m} - \frac{\partial}{\partial t}(\rho_s C_v)\,dx\,dy\,dz$$

$$+ \rho_s C_v v_x \,dy\,dz - \left[\rho_s C_v v_x + \frac{\partial}{\partial x}(\rho_s C_v v_x)\,dx\right] dy\,dz$$

$$+ \rho_s C_v v_y \,dx\,dz - \left[\rho_s C_v v_y + \frac{\partial}{\partial y}(\rho_s C_v v_y)\,dy\right] dx\,dz$$

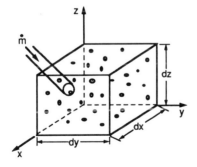

Figure E3.1.1. Differential control volume for sediment continuity

$$+ \rho_s C_v v_z \, dx \, dy - \left[\rho_s C_v v_z + \frac{\partial}{\partial z}(\rho_s C_v v_z) \, dz \right] dx \, dy$$

$$\frac{\partial}{\partial t}(\rho_s C_v) + \frac{\partial}{\partial x}(\rho_s C_v v_x) + \frac{\partial}{\partial y}(\rho_s C_v v_y) + \frac{\partial}{\partial z}(\rho_s C_v v_z) = \frac{\dot{m}}{dx \, dy \, dz}$$

which for constant mass density of sediment ρ_s reduces to

$$\frac{\partial C_v}{\partial t} + \frac{\partial (C_v v_x)}{\partial x} + \frac{\partial (C_v v_y)}{\partial y} + \frac{\partial (C_v v_z)}{\partial z} = \dot{C}_v \qquad \text{(E3.1.1)}$$

where $\dot{C}_v = \dot{\mathbf{V}}_s / \mathbf{V}_t = \dot{m}/\rho_s \mathbf{V}_t$ is the volumetric source of sediment per unit time, including diffusive and mixing fluxes.

The equations governing the conservation of solid mass are thus obtained after substituting ρ_m in Equations (3.10a–c) by the volumetric sediment concentration C_v. A formulation that considers diffusive and mixing fluxes is presented in Section 10.2.

3.2.2 Integral continuity equation

The integral form of the continuity equation is simply the integral of the differential form [Eq. (3.10a)] over a control volume $\mathbf{V}$. For an incompressible fluid, the integral form of conservation of mass is

$$\int_{\mathbf{V}} \frac{\partial \rho_m}{\partial t} \, d\mathbf{V} + \int_{\mathbf{V}} \left(\frac{\partial \rho_m v_x}{\partial x} + \frac{\partial \rho_m v_y}{\partial y} + \frac{\partial \rho_m v_z}{\partial z} \right) d\mathbf{V} = 0 \qquad \text{(3.10e)}$$

This volume integral of velocity gradients can be transformed into surface integrals owing to the divergence theorem applied to vector $\mathbf{F}$

$$\int_{\mathbf{V}} \frac{\partial \mathbf{F}}{\partial x} \, d\mathbf{V} = \int_A F \frac{\partial x}{\partial n} \, dA \qquad \text{(3.11)}$$

in which $\partial x/\partial n$ is the cosine of the angle between the coordinate x and the normal vector $\bar{n}$ pointing outside of the control volume. Example 3.2 illustrates how the integral continuity equation can be directly applied to open channels.

Example 3.2   Application of the integral continuity equation to open channels. Consider the impervious rectangular channel of length X_c sketched in Figure E3.2.1. The differential continuity equation [Eq. (3.10a)] is integrated over the control volume $\mathbf{V} = WhX_c$:

$$\int_{\mathbf{V}} \frac{\partial \rho_m}{\partial t} \, d\mathbf{V} + \int_{\mathbf{V}} \left(\frac{\partial \rho_m v_x}{\partial x} + \frac{\partial \rho_m v_y}{\partial y} + \frac{\partial \rho_m v_z}{\partial z} \right) d\mathbf{V} = 0$$

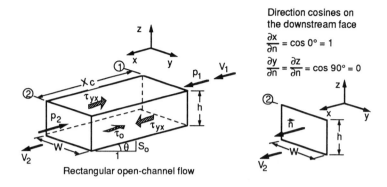

Figure E3.2.1. Rectangular open-channel flow

The first integral is zero for incompressible fluids, while the divergence theorem [Eq. (3.11)] is applied to the second term:

$$\int_A \left(\rho_m v_x \frac{\partial x}{\partial n} + \rho_m v_y \frac{\partial y}{\partial n} + \rho_m v_z \frac{\partial z}{\partial n} \right) dA = 0$$

The values of $\partial x / \partial n$, $\partial y / \partial n$, and $\partial z / \partial n$ are the cosines of the angle between the vector normal to the surface, $\bar{n}$, pointing outside the control volume and the Cartesian coordinates x, y, and z, respectively. Figure E3.2.1 illustrates the direction cosines on the downstream face. Thus, for an incompressible homogeneous suspension $\rho_{m1} = \rho_{m2} = \rho_m$ with the vertical velocity $v_z = dh/dt$ and $v_y = 0$, one obtains

$$A_2 V_2 - A_1 V_1 + W X_c \frac{dh}{dt} = 0$$

which for steady flow, $dh/dt = 0$, reduces to a constant discharge Q:

$$Q = A_1 V_1 = A_2 V_2 \qquad \text{(E3.2.1)} \blacklozenge$$

This integral continuity equation is applicable to steady impervious open-channel flows of incompressible nonsettling homogeneous suspensions. To determine the discharge for complex channel geometries, follow the procedure given in Problem 3.2.

3.3   Equations of motion

The forces acting on a Cartesian element of fluid and sediment ($dx\,dy\,dz$) are classified as either internal forces or external forces. The internal

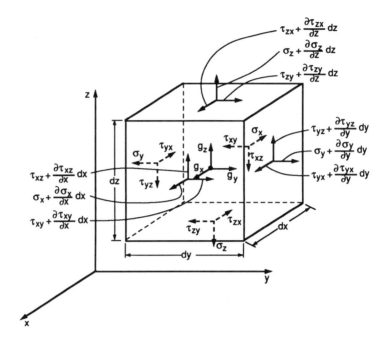

Figure 3.4. Surface stresses on a fluid element

accelerations, or body forces per unit mass, acting at the center of mass of the element are denoted g_x, g_y, and g_z. The external forces per unit area applied on each face of the element are subdivided into normal and tangential stress components. The normal stresses σ_x, σ_y, and σ_z, are designated positive for tension. Six shear stresses, τ_{xy}, τ_{yx}, τ_{xz}, τ_{zx}, τ_{yz}, τ_{zy}, with two orthogonal components on each face are applied, as shown in Figure 3.4. The first letter in each subscript indicates the direction normal to the face and the second designates the direction of the applied stress component. The following identities result from the sum of moments of shear stresses around the centroid: $\tau_{xy} = \tau_{yx}$; $\tau_{xz} = \tau_{zx}$; $\tau_{yz} = \tau_{zy}$.

The cubic element in Figure 3.4 is considered in equilibrium when the sum of the forces per unit mass in each direction x, y, and z equals the corresponding Cartesian acceleration component a_x, a_y, and a_z:

$$a_x = g_x + \frac{1}{\rho_m}\frac{\partial \sigma_x}{\partial x} + \frac{1}{\rho_m}\frac{\partial \tau_{yx}}{\partial y} + \frac{1}{\rho_m}\frac{\partial \tau_{zx}}{\partial z} \tag{3.12a}$$

$$a_y = g_y + \frac{1}{\rho_m}\frac{\partial \sigma_y}{\partial y} + \frac{1}{\rho_m}\frac{\partial \tau_{xy}}{\partial x} + \frac{1}{\rho_m}\frac{\partial \tau_{zy}}{\partial z} \tag{3.12b}$$

$$a_z = g_z + \frac{1}{\rho_m}\frac{\partial \sigma_z}{\partial z} + \frac{1}{\rho_m}\frac{\partial \tau_{xz}}{\partial x} + \frac{1}{\rho_m}\frac{\partial \tau_{yz}}{\partial y} \tag{3.12c}$$

These equations of motion are general and without any restriction as to compressibility, viscous shear, turbulence, or other effects.

The normal stresses can be rewritten as a function of the pressure p and additional normal stresses $\tau_{xx}, \tau_{yy}, \tau_{zz}$ accompanying deformation:

$$\sigma_x = -p + \tau_{xx} \tag{3.13a}$$

$$\sigma_y = -p + \tau_{yy} \tag{3.13b}$$

$$\sigma_z = -p + \tau_{zz} \tag{3.13c}$$

After expanding the acceleration components a_x, a_y, and a_z from Equation (3.12), one can write the equations of motion in Cartesian, cylindical, and spherical coordinates as shown in Table 3.1.

3.4   Euler equations

The Euler equations are simplified forms of the equations of motion [Eqs. (3.14)–(3.16), Table 3.1] for frictionless fluids. In a frictionless fluid, the shear stress components due to deformation are zero (all stress components in $\tau = 0$) and the normal stresses are equal and opposite to the pressure ($\sigma_x = \sigma_y = \sigma_z = -p$). Substitution into the equations of motion yields, for Cartesian coordinates,

$$a_x = g_x - \frac{1}{\rho_m}\frac{\partial p}{\partial x} \tag{3.17a}$$

$$a_y = g_y - \frac{1}{\rho_m}\frac{\partial p}{\partial y} \tag{3.17b}$$

$$a_z = g_z - \frac{1}{\rho_m}\frac{\partial p}{\partial z} \tag{3.17c}$$

These equations, valid for inviscid fluids, are known as the Euler equations.

A gravitation potential $\Omega_g = -g\hat{z}$ can be defined when the axis $\hat{z}$ is vertical upward such that the body acceleration components due to gravity are

$$g_x = \frac{\partial \Omega_g}{\partial x}; \qquad g_y = \frac{\partial \Omega_g}{\partial y}; \qquad g_z = \frac{\partial \Omega_g}{\partial z} \tag{3.18}$$

An example of an application of the Euler equations is the buoyancy force resulting from the integration of the hydrostatic ($a = 0$) pressure distribution around a submerged sphere (see Example 3.3).

Table 3.1. *Equations of motion*

In Cartesian coordinates

x component

$$a_x = \frac{\partial v_x}{\partial t} + v_x\frac{\partial v_x}{\partial x} + v_y\frac{\partial v_x}{\partial y} + v_z\frac{\partial v_x}{\partial z} = g_x - \frac{1}{\rho_m}\frac{\partial p}{\partial x} + \frac{1}{\rho_m}\left(\frac{\partial \tau_{xx}}{\partial x} + \frac{\partial \tau_{yx}}{\partial y} + \frac{\partial \tau_{zx}}{\partial z}\right) \quad (3.14a)$$

y component

$$a_y = \frac{\partial v_y}{\partial t} + v_x\frac{\partial v_y}{\partial x} + v_y\frac{\partial v_y}{\partial y} + v_z\frac{\partial v_y}{\partial z} = g_y - \frac{1}{\rho_m}\frac{\partial p}{\partial y} + \frac{1}{\rho_m}\left(\frac{\partial \tau_{xy}}{\partial x} + \frac{\partial \tau_{yy}}{\partial y} + \frac{\partial \tau_{zy}}{\partial z}\right) \quad (3.14b)$$

z component

$$a_z = \frac{\partial v_z}{\partial t} + v_x\frac{\partial v_z}{\partial x} + v_y\frac{\partial v_z}{\partial y} + v_z\frac{\partial v_z}{\partial z} = g_z - \frac{1}{\rho_m}\frac{\partial p}{\partial z} + \frac{1}{\rho_m}\left(\frac{\partial \tau_{xz}}{\partial x} + \frac{\partial \tau_{yz}}{\partial y} + \frac{\partial \tau_{zz}}{\partial z}\right) \quad (3.14c)$$

In cylindical coordinates

r component

$$\frac{\partial v_r}{\partial t} + v_r\frac{\partial v_r}{\partial r} + \frac{v_\theta}{r}\frac{\partial v_r}{\partial \theta} - \frac{v_\theta^2}{r} + v_z\frac{\partial v_r}{\partial z} = g_r - \frac{1}{\rho_m}\frac{\partial p}{\partial r} + \frac{1}{\rho_m}\left(\frac{1}{r}\frac{\partial}{\partial r}(r\tau_{rr}) + \frac{1}{r}\frac{\partial \tau_{\theta r}}{\partial \theta} - \frac{\tau_{\theta\theta}}{r} + \frac{\partial \tau_{zr}}{\partial z}\right)$$

$$(3.15a)$$

θ component

$$\frac{\partial v_\theta}{\partial t} + v_r\frac{\partial v_\theta}{\partial r} + \frac{v_\theta}{r}\frac{\partial v_\theta}{\partial \theta} + \frac{v_r v_\theta}{r} + v_z\frac{\partial v_\theta}{\partial z} = g_\theta - \frac{1}{\rho_m r}\frac{\partial p}{\partial \theta} + \frac{1}{\rho_m}\left(\frac{1}{r^2}\frac{\partial}{\partial r}(r^2\tau_{r\theta}) + \frac{1}{r}\frac{\partial \tau_{\theta\theta}}{\partial \theta} + \frac{\partial \tau_{z\theta}}{\partial z}\right)$$

$$(3.15b)$$

z component

$$\frac{\partial v_z}{\partial t} + v_r\frac{\partial v_z}{\partial r} + \frac{v_\theta}{r}\frac{\partial v_z}{\partial \theta} + v_z\frac{\partial v_z}{\partial z} = g_z - \frac{1}{\rho_m}\frac{\partial p}{\partial z} + \frac{1}{\rho_m}\left(\frac{1}{r}\frac{\partial(r\tau_{rz})}{\partial r} + \frac{1}{r}\frac{\partial \tau_{\theta z}}{\partial \theta} + \frac{\partial \tau_{zz}}{\partial z}\right) \quad (3.15c)$$

In spherical coordinates

r component

$$\frac{\partial v_r}{\partial t} + v_r\frac{\partial v_r}{\partial r} + \frac{v_\theta}{r}\frac{\partial v_r}{\partial \theta} + \frac{v_\varphi}{r\sin\theta}\frac{\partial v_r}{\partial \varphi} - \frac{v_\theta^2 + v_\varphi^2}{r}$$

$$= g_r - \frac{1}{\rho_m}\frac{\partial p}{\partial r} + \frac{1}{\rho_m}\left(\frac{1}{r^2}\frac{\partial}{\partial r}(r^2\tau_{rr}) + \frac{1}{r\sin\theta}\frac{\partial}{\partial \theta}(\tau_{r\theta}\sin\theta) + \frac{1}{r\sin\theta}\frac{\partial \tau_{r\varphi}}{\partial \varphi} - \frac{\tau_{\theta\theta} + \tau_{\varphi\varphi}}{r}\right) \quad (3.16a)$$

θ component

$$\frac{\partial v_\theta}{\partial t} + v_r\frac{\partial v_\theta}{\partial r} + \frac{v_\theta}{r}\frac{\partial v_\theta}{\partial \theta} + \frac{v_\varphi}{r\sin\theta}\frac{\partial v_\theta}{\partial \varphi} + \frac{v_r v_\theta}{r} - \frac{v_\varphi^2\cot\theta}{r}$$

$$= g_\theta - \frac{1}{\rho_m r}\frac{\partial p}{\partial \theta} + \frac{1}{\rho_m}\left(\frac{1}{r^2}\frac{\partial}{\partial r}(r^2\tau_{r\theta}) + \frac{1}{r\sin\theta}\frac{\partial}{\partial \theta}(\tau_{\theta\theta}\sin\theta) + \frac{1}{r\sin\theta}\frac{\partial \tau_{\theta\varphi}}{\partial \varphi} + \frac{\tau_{r\theta}}{r} - \frac{\tau_{\varphi\varphi}\cot\theta}{r}\right)$$

$$(3.16b)$$

φ component

$$\frac{\partial v_\varphi}{\partial t} + v_r\frac{\partial v_\varphi}{\partial r} + \frac{v_\theta}{r}\frac{\partial v_\varphi}{\partial \theta} + \frac{v_\varphi}{r\sin\theta}\frac{\partial v_\varphi}{\partial \varphi} + \frac{v_\varphi v_r}{r} + \frac{v_\theta v_\varphi}{r}\cot\theta$$

$$= g_\varphi - \frac{1}{\rho_m r\sin\theta}\frac{\partial p}{\partial \varphi} + \frac{1}{\rho_m}\left(\frac{1}{r^2}\frac{\partial}{\partial r}(r^2\tau_{r\varphi}) + \frac{1}{r}\frac{\partial \tau_{\theta\varphi}}{\partial \theta} + \frac{1}{r\sin\theta}\frac{\partial \tau_{\varphi\varphi}}{\partial \varphi} + \frac{\tau_{r\varphi}}{r} + \frac{2\tau_{\theta\varphi}\cot\theta}{r}\right) \quad (3.16c)$$

Example 3.3   Application to buoyancy force on a sphere. Consider the hydrostatic pressure distribution around a sphere of radius R submerged in a fluid of mass density ρ_m (Fig. E3.3.1):

$$p = p_0 - \rho_m gR\cos\theta$$

The buoyancy force F_B is calculated from the volume integral of the pressure gradient from Equation (3.17c), which reduces to the surface integral of the pressure owing to the divergence theorem,

$$F_B = \int_\mathbf{V} \rho_m \left(\frac{-1}{\rho_m}\frac{\partial p}{\partial z}\right) d\mathbf{V} = \int_A -p\frac{\partial z}{\partial n}\, dA$$

where

$$\frac{\partial z}{\partial n} = \cos\theta$$

$$F_B = -\int_A p\cos\theta\, dA$$

The elementary surface area is $dA = (R\sin\theta\, d\varphi)(R\, d\theta)$ and the integration is performed for $0 < \varphi < 2\pi$ and $0 < \theta < \pi$:

$$F_B = -\int_0^\pi \int_0^{2\pi} p\cos\theta(R\sin\theta\, d\varphi)(R\, d\theta)$$

$$F_B = 0 + 2\pi\rho_m gR^3 \int_0^\pi \cos^2\theta\sin\theta\, d\theta = 2\pi\gamma_m R^3 \times \frac{2}{3}$$

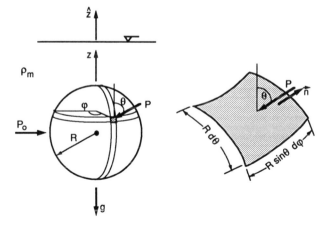

Figure E3.3.1. Buoyancy force on a sphere

We learn that the integral of a constant on a closed surface is zero. The hydrostatic pressure distribution therefore gives the following buoyancy force:

$$F_B = \frac{4\pi}{3}\gamma_m R^3 = \gamma_m \forall_{sphere}$$

(E3.3.1) ◆

Notice that the result is the same as the weight of displaced fluid $F_B = \int_\forall \rho_m(-g_z)\,d\forall$ from Equation (3.17c), because $a_z = 0$.

3.5   Bernoulli equation

The Bernoulli equation also represents a particular form of the equations of motion for steady irrotational flow of frictionless fluids. After considering the equations of motion [Eqs. (3.9) and (3.14)] and the gravitation potential [Eq. (3.18)], we can rewrite the equations of motion for incompressible sediment-laden fluids of mass density ρ_m as

$$\frac{\partial}{\partial x}\left(\frac{p}{\rho_m} - \Omega_g + \frac{v^2}{2}\right) = (v_y \otimes_z - v_z \otimes_y) - \frac{\partial v_x}{\partial t}$$

$$+ \frac{1}{\rho_m}\left(\frac{\partial \tau_{xx}}{\partial x} + \frac{\partial \tau_{yx}}{\partial y} + \frac{\partial \tau_{zx}}{\partial z}\right)$$

(3.19a)

$$\frac{\partial}{\partial y}\left(\frac{p}{\rho_m} - \Omega_g + \frac{v^2}{2}\right) = (v_z \otimes_x - v_x \otimes_z) - \frac{\partial v_y}{\partial t}$$

$$+ \frac{1}{\rho_m}\left(\frac{\partial \tau_{xy}}{\partial x} + \frac{\partial \tau_{yy}}{\partial y} + \frac{\partial \tau_{zy}}{\partial z}\right)$$

(3.19b)

$$\frac{\partial}{\partial z}\left(\frac{p}{\rho_m} - \Omega_g + \frac{v^2}{2}\right) = (v_x \otimes_y - v_y \otimes_x) - \frac{\partial v_z}{\partial t}$$

$$+ \frac{1}{\rho_m}\left(\frac{\partial \tau_{xz}}{\partial x} + \frac{\partial \tau_{yz}}{\partial y} + \frac{\partial \tau_{zz}}{\partial z}\right)$$

(3.19c)

For steady irrotational flow of frictionless fluids, the right-hand side of Equations (3.19) vanishes and the Bernoulli sum H for homogeneous incompressible fluids is constant throughout the fluid:

$$H = \frac{p}{\gamma_m} + \hat{z} + \frac{v^2}{2g} = ct$$

(3.20) ◆

In the particular case of flow in a horizontal plane (constant $\hat{z}$) of a homogeneous fluid (constant ρ_m), the pressure at any point p where the velocity v can be calculated from the pressure p_r at any reference point given the reference velocity v_r:

$$p = \frac{\rho_m}{2}(v_r^2 - v^2) + p_r \tag{3.21}$$

When the flow is steady, frictionless, but rotational, the right-hand side of Equation (3.19) equals zero only along a streamline [because of Eq. (3.7)]; hence, for a homogeneous incompressible fluid,

$$H = \frac{p}{\gamma_m} + \hat{z} + \frac{v^2}{2g} = ct \quad \text{along a streamline} \tag{3.20a}$$

For the particular case of unsteady irrotational flow of homogeneous incompressible fluids,

$$\frac{1}{g}\frac{\partial v_x}{\partial t} = -\frac{\partial H}{\partial x} \tag{3.22a}$$

$$\frac{1}{g}\frac{\partial v_y}{\partial t} = -\frac{\partial H}{\partial y} \tag{3.22b}$$

$$\frac{1}{g}\frac{\partial v_z}{\partial t} = -\frac{\partial H}{\partial z} \tag{3.22c}$$

These equations remain applicable only when all shear stresses are negligible.

For steady flow at a mean flow velocity V in a wide rectangular channel at a bed slope θ, the first and third terms of the Bernoulli sum describe the specific energy function E defined as

$$E = \frac{p}{\gamma_m} + \frac{V^2}{2g} = h\cos^2\theta + \frac{V^2}{2g} \tag{3.23a}$$

Considering the unit discharge $q = Vh$ and the Froude number Fr defined as $\text{Fr}^2 = V^2/gh = q^2/gh^3$, the function E as $\theta \to 0$ is approximated by

$$E = h + \frac{q^2}{2gh^2} = h\left(1 + \frac{\text{Fr}^2}{2}\right) \tag{3.23b}$$

The properties of the function E are such that, under constant unit discharge q, the critical depth h_c and the critical velocity V_c correspond to the minimum ($\partial E/\partial h = 0$) of Equation (3.23b). The minimum value E_{min} is found when $\text{Fr}_c^2 = V_c^2/gh_c = q^2/gh_c^3 = 1$.

One can easily demonstrate that $q^2 = gh_c^3$ or $h_c = \sqrt[3]{q^2/g}$, and $E_{min} = 1.5h_c$, resulting in the following identities for the Froude number:

$$\text{Fr} = \frac{V}{\sqrt{gh}} = \frac{q}{h\sqrt{gh}} = \left(\frac{h_c}{h}\right)^{3/2} \tag{3.24}$$

The Froude number is an approximate indicator of the ratio of flow velocity V to surface wave celerity $c_w \cong \sqrt{gh}$ in shallow open channels. An open-channel flow application of the Bernoulli equation is outlined in Example 3.4.

Example 3.4 Application to accelerating open-channel flow. Consider steady flow in a wide rectangular channel. Determine

(a) the maximum possible elevation of a sill Δz at section A that will not cause backwater; and
(b) the maximum lateral contraction of the channel at section A that will not cause backwater.

The accelerating flow is shown with the specific energy diagram in Figure E3.4.1.

(a) The maximum elevation of the sill Δz_{max} at section A is such that the flow will be critical on top of the sill and $\Delta z_{max} + E_{min} = E_1$, or $\Delta z_{max} = \Delta E$;

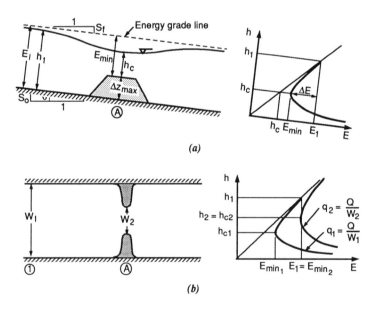

Figure E3.4.1. Flow near (a) sill and (b) abutment

(b) the minimum channel width W_2 at section A that will not cause backwater is such that the total discharge remains constant, $Q = W_1 q_1 = W_2 q_2$, and the flow is critical in the contracted section A_2, or $h_{c2} = 0.67E_1 = 0.67E_{min_2}$; and $Fr_{c2} = 1 = q_2^2/gh_{c2}^3$, or $W_2 = Q/\sqrt{g(0.67E_1)^3}$.

3.6   Momentum equations

Momentum equations define the hydrodynamic forces exerted by sediment-laden flows. After the equations of motion [Eqs. (3.14)–(3.16)] are multiplied by the mass density of the mixture ρ_m, the terms on the left-hand side of the equations represent the rate of momentum change per unit volume, while the rate of impulse per unit volume is found on the right-hand side. Integration over the total volume $\forall$ shows that the rate of momentum change equals the impulse per unit time. For example, the x component in the Cartesian coordinates is

$$\int_\forall \rho_m\left(\frac{\partial v_x}{\partial t} + v_x\frac{\partial v_x}{\partial x} + v_y\frac{\partial v_x}{\partial y} + v_z\frac{\partial v_x}{\partial z}\right)d\forall$$

$$= \int_\forall \rho_m g_x\, d\forall - \int_\forall \frac{\partial p}{\partial x}\, d\forall + \int_\forall \left(\frac{\partial \tau_{xx}}{\partial x} + \frac{\partial \tau_{yx}}{\partial y} + \frac{\partial \tau_{zx}}{\partial z}\right)d\forall \qquad (3.25)$$

The integrand on the left-hand side can be rewritten as follows:

$$\frac{\partial \rho_m v_x}{\partial t} + \frac{\partial \rho_m v_x^2}{\partial x} + \frac{\partial \rho_m v_x v_y}{\partial y} + \frac{\partial \rho_m v_x v_z}{\partial z}$$

$$- v_x\left(\frac{\partial \rho_m}{\partial t} + \frac{\partial \rho_m v_x}{\partial x} + \frac{\partial \rho_m v_y}{\partial y} + \frac{\partial \rho_m v_z}{\partial z}\right) \qquad (3.26)$$

By virtue of the continuity equation [Eq. (3.10a)], the terms in parentheses in Equation (3.26) can be dropped. Hence, the integral of the first partial derivative is equal to the total derivative of a volume integral. The volume integral of the remaining momentum and stress terms can be transformed into surface integrals by means of the divergence theorem [Eq. (3.11)]. The result is the general impulse–momentum relationship.

x component

$$\frac{d}{dt}\int_\forall \rho_m v_x\, d\forall + \int_A \rho_m v_x\left(v_x\frac{\partial x}{\partial n} + v_y\frac{\partial y}{\partial n} + v_z\frac{\partial z}{\partial n}\right)dA$$

$$= \int_\forall \rho_m g_x\, d\forall - \int_A p\frac{\partial x}{\partial n}\, dA + \int_A \left(\tau_{xx}\frac{\partial x}{\partial n} + \tau_{yx}\frac{\partial y}{\partial n} + \tau_{zx}\frac{\partial z}{\partial n}\right)dA \qquad (3.27a)$$

y component

$$\frac{d}{dt}\int_{\Psi}\rho_{m}v_{y}\,d\Psi+\int_{A}\rho_{m}v_{y}\left(v_{x}\frac{\partial x}{\partial n}+v_{y}\frac{\partial y}{\partial n}+v_{z}\frac{\partial z}{\partial n}\right)dA$$

$$=\int_{\Psi}\rho_{m}g_{y}\,d\Psi-\int_{A}p\frac{\partial y}{\partial n}\,dA+\int_{A}\left(\tau_{xy}\frac{\partial x}{\partial n}+\tau_{yy}\frac{\partial y}{\partial n}+\tau_{zy}\frac{\partial z}{\partial n}\right)dA \quad (3.27b)$$

z component

$$\frac{d}{dt}\int_{\Psi}\rho_{m}v_{z}\,d\Psi+\int_{A}\rho_{m}v_{z}\left(v_{x}\frac{\partial x}{\partial n}+v_{y}\frac{\partial y}{\partial n}+v_{z}\frac{\partial z}{\partial n}\right)dA$$

$$=\int_{\Psi}\rho_{m}g_{z}\,d\Psi-\int_{A}p\frac{\partial z}{\partial n}\,dA+\int_{A}\left(\tau_{xz}\frac{\partial x}{\partial n}+\tau_{yz}\frac{\partial y}{\partial n}+\tau_{zz}\frac{\partial z}{\partial n}\right)dA \quad (3.27c)$$

It is observed that momentum is a vector quantity, the momentum change due to convection is embodied in the surface integral on the left-hand side of Equation (3.27), and all the stresses are expressed in terms of surface integrals. Example 3.5 provides a detailed application of the momentum equations to open-channel flows.

Example 3.5   Analysis of momentum equations for open channels. With reference to the rectangular channel sketched in Figure E3.5.1, the momentum relationship [Eq. (3.27a)] in the downstream x direction is applied to an open channel, now subjected to rainfall at an angle θ_r, velocity V_r over an area A_r, wind shear τ_w, bank shear $\tau_s = \tau_{yx}$, and bed shear $\tau_b = \tau_0 = \tau_{zx}$:

$$\frac{d}{dt}\int_{\Psi}\rho_{m}v_{x}\,d\Psi+\int_{A}\rho_{m}v_{x}\left(v_{x}\frac{\partial x}{\partial n}+v_{y}\frac{\partial y}{\partial n}+v_{z}\frac{\partial z}{\partial n}\right)dA$$

$$=\int_{\Psi}\rho_{m}g_{x}\,d\Psi-\int_{A}p\frac{\partial x}{\partial n}\,dA+\int_{A}\left(\tau_{xx}\frac{\partial x}{\partial n}+\tau_{yx}\frac{\partial y}{\partial n}+\tau_{zx}\frac{\partial z}{\partial n}\right)dA$$

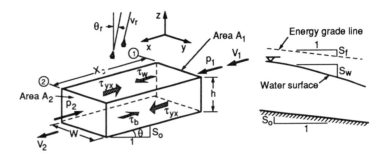

Figure E3.5.1. Momentum equations in open channels

The first integral can be dropped for steady flow. Others vanish for one-dimensional flow in impervious channels, $v_y = v_z = \tau_{xx} = 0$, leaving

$$\int_A \rho_m v_x^2 \frac{\partial x}{\partial n}\, dA + \int_A p \frac{\partial x}{\partial n}\, dA = \int_\forall \rho_m g_x\, d\forall + \int_A \tau_{zx} \frac{\partial z}{\partial n}\, dA + \int_A \tau_{yx} \frac{\partial y}{\partial n}\, dA$$

Consider an incompressible homogeneous fluid, $\rho_m = ct$, and define the momentum correction factor β_m given the cross-section-averaged velocity V_x:

$$\beta_m = \frac{1}{AV_x^2} \int_A v_x^2\, dA \qquad\qquad \text{(E3.5.1)} \blacklozenge$$

The value of β_m is generally close to unity; the reader can refer to Example 6.1 for a detailed calculation example. With average values of pressure p, velocity V, and area A at the upstream cross section 1 and downstream cross section 2, the integration of the momentum equation for this control volume $\forall$ of length X_c, width W, and height h yields

$$\beta_m \rho_m A_2 V_2^2 + p_2 A_2 - \beta_m \rho_m A_1 V_1^2 - p_1 A_1 - \rho A_r V_r^2 \sin(\theta + \theta_r)$$
$$= \gamma_m \forall \sin\theta - \tau_b W X_c - \tau_s 2h X_c + \tau_w W X_c$$

Assuming that the boundary shear stress τ_0 equals the bank shear stress τ_s and the bed shear stress τ_b, the equation with negligible rainfall, $A_r \to 0$, without wind shear, $\tau_w \to 0$, can be rewritten when the channel inclination θ is small ($\sin\theta \cong S_0$ the bed slope) as

$$p_2 A_2 + \beta_m \rho_m A_2 V_2^2 + \tau_0 (W + 2h) X_c$$
$$= p_1 A_1 + \beta_m \rho_m A_1 V_1^2 + \gamma_m \left(\frac{A_1 + A_2}{2} \right) X_c S_0$$

Further reduction of this equation is possible for uniform flow ($A = A_1 = A_2$), in which case $p_1 A_1 = p_2 A_2$, $\beta_m \rho_m V_1^2 = \beta_m \rho_m V_2^2$ and the friction slope S_f equals both the water surface slope S_w and the bed slope S_0. The boundary shear stress τ_0 is related to the friction slope S_f in the following manner:

$$\tau_0 = \gamma_m \frac{A}{W + 2h} S_f = \gamma_m \frac{A}{P} S_f = \gamma_m R_h S_f \qquad \text{(E3.5.2a)} \blacklozenge$$

where the hydraulic radius $R_h = A/P$ is the ratio of the cross-sectional area $A = Wh$ to the wetted perimeter $P = W + 2h$. For wide rectangular channels, $W \gg h$ and $R_h = h$, this equation reduces to $\tau_0 = \gamma_m h S_f$. For the particular case of steady uniform flow,

$$\tau_0 = \gamma_m R_h S_0 \quad \text{when } S_f = S_0 \qquad\qquad \text{(E3.5.2b)}$$

After defining the Darcy–Weisbach friction factor $f = 8\tau_0/\rho_m V^2$, one obtains $(f/8)\rho_m V^2 = \rho_m g R_h S_f$, or

$$V = \sqrt{\frac{8g}{f}} \, R_h^{1/2} S_f^{1/2} \qquad (E3.5.3) \blacklozenge$$

For steady uniform flow in a wide rectangular channel, $R_h = h$, the velocity in Equation (E3.5.3) can be replaced by the unit discharge $q = Vh$ to give the normal depth $h_n = (fq^2/8gS_f)^{1/3} = (f/8g)(V^2/S_f)$.

For steady uniform flow, $S_0 = S_f$, the flow depth reaches the normal flow depth h_n, or $h_n = (fq^2/8gS_0)^{1/3} = (f/8g)(V_n^2/S_0)$.

From the ratio of h to h_n at constant unit discharge q and friction factor f, one obtains

$$\frac{S_f}{S_0} = \left(\frac{h_n}{h}\right)^3 \qquad (E3.5.4) \blacklozenge$$

3.7   Energy equation

The rate of work done by fluid motion within a control volume $\forall$ is obtained by integrating the product of the equations of force per unit volume, $\rho_m a_x, \rho_m a_y, \rho_m a_z$ [from Eq. (3.12)], and the velocity component v_x, v_y, v_z in the same direction. The Cartesian components in the three orthogonal directions are then added to give a single scalar equation of rate of work done, or power:

$$\int_\forall \rho_m (a_x v_x + a_y v_y + a_z v_z)\, d\forall$$

$$= \int_\forall \rho_m (v_x g_x + v_y g_y + v_z g_z)\, d\forall$$

$$+ \int_\forall \left[v_x \left(\frac{\partial \sigma_x}{\partial x} + \frac{\partial \tau_{yx}}{\partial y} + \frac{\partial \tau_{zx}}{\partial z} \right) + v_y \left(\frac{\partial \tau_{xy}}{\partial x} + \frac{\partial \sigma_y}{\partial y} + \frac{\partial \tau_{zy}}{\partial z} \right) \right.$$

$$\left. + v_z \left(\frac{\partial \tau_{xz}}{\partial x} + \frac{\partial \tau_{yz}}{\partial y} + \frac{\partial \sigma_z}{\partial z} \right) \right] d\forall \qquad (3.28)$$

The vorticity components in the equations of motion [Eq. (3.9)] cancel out and the left-hand side of Equation (3.28) can be written solely as a function of the square of the velocity magnitude v^2. The body forces are derived from the gravitation potential Ω_g, which is time invariant. The elastic energy of the fluid per unit mass Ω_e is defined as follows:

$$\rho \frac{d\Omega_e}{dt} = -p(\Theta_x + \Theta_y + \Theta_z) = -p\left(\frac{\partial v_x}{\partial x} + \frac{\partial v_y}{\partial y} + \frac{\partial v_z}{\partial z}\right) \qquad (3.29)$$

This elastic energy Ω_e vanishes for incompressible fluids.

Using the divergence theorem, the equation of power [Eq. (3.28)] is transformed to

$$\frac{d}{dt} \int_{\Psi} \rho_m \left(\frac{v^2}{2} - \Omega_g + \Omega_e\right) d\Psi$$

$$+ \int_A \rho_m \left(\frac{v^2}{2} - \Omega_g + \Omega_e\right) \left(v_x \frac{\partial x}{\partial n} + v_y \frac{\partial y}{\partial n} + v_z \frac{\partial z}{\partial n}\right) dA$$

$$= \int_A \left[(v_x \sigma_x + v_y \tau_{xy} + v_z \tau_{xz}) \frac{\partial x}{\partial n} + (v_x \tau_{yx} + v_y \sigma_y + v_z \tau_{yz}) \frac{\partial y}{\partial n} \right.$$

$$+ (v_x \tau_{zx} + v_y \tau_{zy} + v_z \sigma_z) \frac{\partial z}{\partial n} \bigg] dA$$

$$- \int_{\Psi} \left[\tau_{xx} \frac{\partial v_x}{\partial x} + \tau_{yy} \frac{\partial v_y}{\partial y} + \tau_{zz} \frac{\partial v_z}{\partial z} + \tau_{xy}(\oslash_z) + \tau_{xz}(\oslash_y) + \tau_{yz}(\oslash_x) \right] d\Psi$$

$$(3.30) \ \blacklozenge$$

The rate of work done is a scalar quantity. Convective terms reduce to the net fluxes across the surfaces. The surface integral on the right-hand side represents the total rate of work done by the external stresses, which is conservative. The volume integral then indicates the rate at which mechanical energy gradually transforms into heat.

Example 3.6 details the application of the power equation to the rectangular open channel considered in Example 3.5. The analysis of backwater curves in Example 3.7 highlights the application of the energy equation to gradually varied flow.

Example 3.6 Analysis of rate of work done in open channels. Consider the application of the power equation [Eq. (3.28)] to the open-channel flow illustrated in Example 3.5 (see Fig. E3.5.1):

$$\frac{d}{dt} \int_{\Psi} \rho_m \left(\frac{v^2}{2} - \Omega_g + \Omega_e\right) d\Psi$$

$$+ \int_A \rho_m \left(\frac{v^2}{2} - \Omega_g + \Omega_e\right) \left(v_x \frac{\partial x}{\partial n} + v_y \frac{\partial y}{\partial n} + v_z \frac{\partial z}{\partial n}\right) dA$$

$$= \int_A \left[(v_x \sigma_x + v_y \tau_{xy} + v_z \tau_{xz}) \frac{\partial x}{\partial n} + (v_x \tau_{yx} + v_y \sigma_y + v_z \tau_{yz}) \frac{\partial y}{\partial n} \right.$$

$$+ (v_x \tau_{zx} + v_y \tau_{zy} + v_z \sigma_z) \frac{\partial z}{\partial n} \right] dA$$

$$- \int_{\Psi} \left[\tau_{xx} \frac{\partial v_x}{\partial x} + \tau_{yy} \frac{\partial v_y}{\partial y} + \tau_{zz} \frac{\partial v_z}{\partial z} \right.$$

$$\left. + \tau_{xy} \left(\frac{\partial v_y}{\partial x} + \frac{\partial v_x}{\partial y} \right) + \tau_{xz} \left(\frac{\partial v_x}{\partial z} + \frac{\partial v_z}{\partial x} \right) + \tau_{yz} \left(\frac{\partial v_z}{\partial y} + \frac{\partial v_y}{\partial z} \right) \right] d\Psi$$

The first integral can be dropped for steady flow. Further simplification arises for incompressible fluids ($\Omega_e = 0$ from the continuity relationship) and one-dimensional flow ($v_y = v_z = 0$). If we assume that bank and transversal shear stresses are negligible, $\tau_{xx} = \tau_{yy} = \tau_{zz} = \tau_{zy} = \tau_{yz} = \tau_{yx} = \tau_{xy} = 0$, only the bed shear stress in the downstream direction is nonzero $\tau_{zx} = \tau_{xz} = \tau_0 = \tau_b \neq 0$, and the energy equation reduces to

$$\int_A \rho_m \left(\frac{v^2}{2} - \Omega_g \right) v_x \frac{\partial x}{\partial n} \, dA$$

$$= \int_A -v_x p \frac{\partial x}{\partial n} \, dA + \int_A v_x \tau_{zx} \frac{\partial z}{\partial n} \, dA - \int_{\Psi} \tau_{xz} \frac{\partial v_x}{\partial z} \, d\Psi$$

The last surface integral vanishes, because v_x equals zero at the bed and τ_{zx} is zero at the free surface.

After combining the first two surface integrals with $\Omega_g = -g\hat{z}$, one obtains the integral form of the energy equation

$$\int_A \rho_m g \left(\frac{v^2}{2g} + \hat{z} + \frac{p}{\gamma_m} \right) v_x \frac{\partial x}{\partial n} \, dA = -\int_{\Psi} \tau_{xz} \frac{\partial v_x}{\partial z} \, d\Psi \qquad \text{(E3.6.1)}$$

For one-dimensional flow, the energy correction factor α_e is defined as

$$\alpha_e = \frac{1}{V^3 A} \int_A v_x^3 \, dA \qquad \text{(E3.6.2)}$$

where V is the cross-section-averaged velocity. A calculation example for α_e, given in Example 6.1, illustrates that numerical values of α_e remain close to unity. The volume integral of Equation E3.6.1 defines the head loss ΔH_L:

$$\Delta H_L = \frac{1}{\gamma_m Q} \int_{\Psi} \tau_{xz} \frac{\partial v_x}{\partial z} \, d\Psi$$

Assuming a hydrostatic pressure distribution, the integral form of the energy equation [Eq. (E3.6.1)] is rewritten as

$$\gamma_m Q\left(\frac{p_2}{\gamma_m}+\hat{z}_2+\alpha_{e_2}\frac{V_2^2}{2g}\right)-\gamma_m Q\left(\frac{p_1}{\gamma_m}+\hat{z}_1+\alpha_{e_1}\frac{V_1^2}{2g}\right)=-\gamma_m Q\,\Delta H_L$$

$$\underbrace{\hphantom{\gamma_m Q\left(\frac{p_2}{\gamma_m}+\hat{z}_2+\alpha_{e_2}\frac{V_2^2}{2g}\right)}}_{\tilde{H}_2}\qquad\underbrace{\hphantom{\gamma_m Q\left(\frac{p_1}{\gamma_m}+\hat{z}_1+\alpha_{e_1}\frac{V_1^2}{2g}\right)}}_{\tilde{H}_1}$$

or

$$\Delta H_L = \tilde{H}_1 - \tilde{H}_2$$

in which the integral form of the Bernoulli sum $\tilde{H}$ resembles the differential formulation [Eq. (3.20)], except for the energy correction factor α_e. This integral form of the Bernoulli equation which includes head loss is appropriate for open-channel flows.

The integral form of the specific energy $\tilde{E}$ follows

$$\frac{\Delta\tilde{H}}{\Delta x}=\frac{\Delta}{\Delta x}\left(\frac{p}{\gamma_m}+\hat{z}+\alpha_e\frac{V^2}{2g}\right)=-S_f$$

$$\frac{\Delta\tilde{E}}{\Delta x}=\frac{\Delta}{\Delta x}\left(\frac{p}{\gamma_m}+\alpha_e\frac{V^2}{2g}\right)=-\frac{\Delta\hat{z}}{\Delta x}-S_f=S_0-S_f \qquad\text{(E3.6.3)} \blacklozenge$$

in which $\tilde{E}=(p/\gamma_m)+(\alpha_e V^2/2g)$ is the integral form of specific energy.

Notice the similarity between $\tilde{E}$ and E from Equation (3.23a), considering that α_e remains close to unity in most turbulent flows.

Example 3.7 Analysis of backwater curves. Water surface elevation profiles commonly called *backwater curves* result from a direct application of the integral form of the energy equation. In the simplified case of steady one-dimensional flow, Equation (E3.6.3) can be rewritten as

$$\frac{d\tilde{E}}{dx}=\frac{d\tilde{E}}{dh}\frac{dh}{dx}=S_0-S_f$$

From Equation (E3.6.3), the derivative of $\tilde{E}$ with respect to h, given that $p=\gamma_m h$ and $q=Vh$, is

$$\left(1-\alpha_e\frac{q^2}{gh^3}\right)\frac{dh}{dx}=S_0-S_f$$

After substituting the Froude number [Eq. (3.24)], and with $\alpha_e\cong 1$, one obtains the relationship describing water surface elevation for steady one-dimensional flow of an incompressible sediment-laden fluid:

$$\frac{dh}{dx}=\frac{S_0-S_f}{1-\mathrm{Fr}^2} \qquad\text{(E3.7.1)} \blacklozenge$$

Using the properties of critical flow depth h_c from Equation (3.24) and normal depth h_n from Equation (E3.5.4), in wide rectangular channels,

$R_h = h$, the governing equation for steady flow, with constant q and f, becomes

$$\frac{dh}{dx} = S_0\left[1-\left(\frac{h_n}{h}\right)^3\right]\Big/\left[1-\left(\frac{h_c}{h}\right)^3\right] \qquad (E3.7.2)$$

where

$$h_n = \left(\frac{fq^2}{8gS_0}\right)^{1/3} \quad \text{and} \quad h_c = \left(\frac{q^2}{g}\right)^{1/3}$$

Notice that $dh/dx \to 0$ as the flow depth h approaches the normal depth h_n. Also, $dh/dx \to \infty$ near critical flow as $h \to h_c$.

The sign of dh/dx depends on the relative magnitudes of h, h_n, and h_c. Five types of backwater profiles are possible:

1. H profiles for horizontal surfaces with $\quad h_n \to \infty$
2. M profiles for mild slopes when $\qquad\quad h_n > h_c$
3. C profiles for critical slopes when $\qquad h_n = h_c$
4. S profiles for steep slopes when $\qquad\; h_n < h_c$
5. A profiles for adverse slopes when $\qquad S_0 < 0$

Typical water surface profiles in downstream-sloping open channels are shown in Figure E3.7.1.

Numerical calculations can be initiated from a given flow depth h_1, and the distance increment Δx at which $h_2 = h_1 \pm \Delta h$ is approximated by

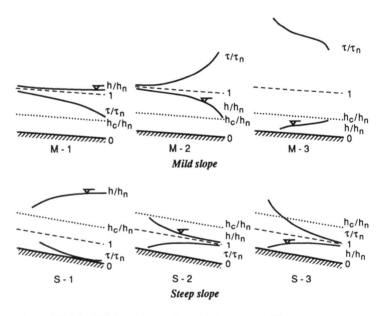

Figure E3.7.1. Mild and steep slope backwater profiles

$$\Delta x \cong \Delta h \left[1 - \left(\frac{h_c}{h_1}\right)^3 \right] \bigg/ S_0 \left[1 - \left(\frac{h_n}{h_1}\right)^3 \right]$$ (E3.7.3)

Notice that it is also possible to iterate on Δh until a predetermined value of Δx is obtained. The friction slope S_f in gradually varied flows with constant q and f can be approximated by Equation (E3.5.4), which shows that $S_f < S_0$ when the flow depth exceeds the normal depth and increases very rapidly as $h < h_n$.

Likewise, a reasonable first approximation for bed shear stress τ_b in gradually varied flows, with constant q and f, is compared with the bed shear stress at normal flow depth τ_{bn}:

$$\frac{\tau_b}{\tau_{bn}} \cong \frac{\gamma_m h S_f}{\gamma_m h_n S_0} = \frac{h}{h_n}\left(\frac{h_n}{h}\right)^3 = \left(\frac{h_n}{h}\right)^2$$ (E3.7.4)

This shows that the bed shear stress increases ($\tau > \tau_n$) at flow depths less than normal depth ($h < h_n$). Bed shear stress distributions for one-dimensional mild M and steep S backwater curves are sketched in Figure E3.7.1. The analysis shows that the shear stress increases in the downstream direction for converging flows (M-2 and S-2 backwater curves) and decreases for diverging flows (M-1, M-3, S-1, and S-3 backwater curves).

Exercises

3.1. Demonstrate, using Equations (3.6) and (3.4), that div $\bar{\omega} = 0$ for homogeneous incompressible fluids.

**3.2. Demonstrate that the equations of motion [Eqs. (3.8a) and (3.9a)] reduce to Equation (3.1a).

3.3. Demonstrate the continuity relationship [Eq. (3.10a)] in Cartesian coordinates by considering the internal mass change and the balance of mass fluxes entering and leaving a cubic control volume element:

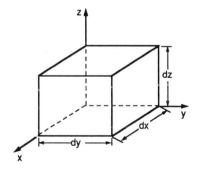

*3.4. Derive the x component of the equation of motion in Cartesian coordinates [Eq. (3.14a)] from the force diagram in Figure 3.4.

3.5. Derive the x component of the Bernoulli equation [Eq. (3.19a)] from Equations (3.9a), (3.14a), and (3.18).

3.6. Derive the x component of the momentum equation [Eq. (3.27a)] from Equation (3.25).

3.7. Demonstrate that the power equation [Eq. (3.30)] stems from Equation (3.28).

*3.8. Demonstrate, from the specific energy function E, that $q^2 = gh_c^3$ and $E_{min} = 3h_c/2$ for steady one-dimensional open-channel flow.

*Problem 3.1

With reference to Figure 3.2, determine which type of deformation is obtained when $v_x = 2y$ and $v_y = v_z = 0$.

Answer: Translation along x only; no linear deformation; angular deformation $\oslash_z = 2$; rotation $\otimes_z = -2$.

**Problem 3.2

Given the 280-ft-wide cross section (depth measurements every 10 ft) and depth-averaged velocity profile below, calculate:

(a) the total cross-sectional area, $A = \Sigma_i a_i$, where a_i is the incremental cross-sectional area;

(b) the total discharge, $Q = \Sigma_i a_i v_i$, where v_i is the depth-averaged flow velocity normal to the incremental area; and

(c) the cross-sectional average flow velocity $V = Q/A$.

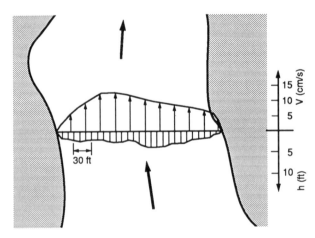

*Problem 3.3

Calculate the magnitude and direction of the buoyancy force applied on a sphere submerged under steady one-dimensional flow ($v_y = v_z = 0$) on a steep slope. Assume that the particle is stationary with respect to the surrounding inviscid fluid of density ρ_m. Compare the results with Example 3.3. [*Hint:* Integrate the pressure distribution around the sphere from Equation (3.17c) with $a_z = 0$.]

Answer: $F_B = \gamma_m \forall \cos 30°$ in direction z, which is less than F_B in Example 3.3. Notice that the sphere will accelerate in the x direction at $a_x = g_x = g \sin 30° = 4.9$ m/s^2 until $(1/\rho_m)(\partial \tau_{yx}/\partial y) = -g_x$. In all cases, the x component of the buoyancy force vanishes.

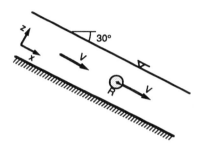

**Computer Problem 3.1

Consider steady flow ($q = 3.72$ m^2/s) in the following impervious rigid boundary channel:

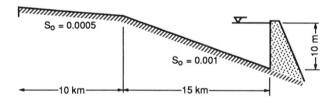

Assume that the channel width remains large and constant regardless of flow depth, and $f = 0.03$. Determine the distribution of the following parameters along the 25-km reach of the channel when the water surface elevation at the dam is 10 m above the bed elevation: (a) flow depth in m; (b) mean flow velocity in m/s; and (c) bed shear stress in N/m^2.

4

Particle motion in inviscid fluids

The analysis of particle motion in inviscid fluids is important because at large particle Reynolds numbers the flow of a viscous fluid around a particle may be regarded as consisting of two regions: (1) a thin layer called the boundary layer adjacent to the body, in which the viscous effects are dominant; and (2) the surrounding fluid, where the viscous effects are negligible. In this outer region the flow of a viscous fluid is essentially that of an ideal fluid and the pressure is transmitted to the boundary through the boundary layer.

The flow of inviscid fluids around submerged particles may be due to either the movement of the fluid or the movement of the particle or a combination of both. The following discussion considers flow conditions made steady by application of the principle of relative motion of the particle with respect to the surrounding fluid. The analysis focuses on simple particle shapes, such as cylinders (Section 4.1) and spheres (Section 4.2), to illustrate the fundamentals with simple calculations.

At every point on the surface of a submerged particle, the fluid exerts a force per unit area or stress. In Chapter 3, the stress vector was divided into a pressure component acting in a direction normal to the surface and two orthogonal shear stress components acting in the plane tangent to the surface. In this chapter, the stress vector for inviscid fluids is always normal to the surface, which means that there is only a pressure component and no shear stress. The flow of inviscid fluids is irrotational and can be described by the potential flow theory. Velocity components are derived from the velocity potential function. The Bernoulli equation applies throughout the flow field, and the pressure can be defined whenever the velocity is obtainable. The drag and lift forces can then be calculated after integrating the pressure distribution around the surface of a body. Streamlines Ψ are associated with potential lines Φ to form orthogonal flow nets, shown in Figure 4.1.

50

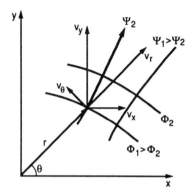

Figure 4.1. Flow net

Steady two-dimensional flow nets are mathematically described by potential functions Φ and stream functions Ψ. The velocity components v_x and v_y are definable from the velocity potential Φ or the stream function Ψ:

$$v_x = -\frac{\partial \Phi}{\partial x} = -\frac{\partial \Psi}{\partial y} \qquad (4.1a)$$

$$v_y = -\frac{\partial \Phi}{\partial y} = \frac{\partial \Psi}{\partial x} \qquad (4.1b)$$

The continuity relationship for incompressible fluids ($\partial v_x/\partial x + \partial v_y/\partial y$ equals zero) satisfies the condition that Ψ give an exact differential expression and Φ obey the Laplace equation ($\nabla^2 \Phi = \partial^2 \Phi/\partial x^2 + \partial^2 \Phi/\partial y^2 = 0$). The stream function is therefore a consequence of the continuity equation and is applicable to both incompressible rotational and irrotational flows.

The potential function Φ gives an exact differential expression only when the vorticity is zero ($\partial v_x/\partial y - \partial v_y/\partial x = 0$). This potential function is thus applicable only to irrotational flows and can be defined by the Laplace equation ($\nabla^2 \Psi = 0$).

In cylindrical coordinates, the velocity components v_r in the radial direction and v_θ in the tangential direction (v_θ is positive in the direction of increasing θ) of a system are

$$v_r = -\frac{\partial \Phi}{\partial r} = -\frac{1}{r}\frac{\partial \Psi}{\partial \theta} \qquad (4.2a)$$

$$v_\theta = -\frac{1}{r}\frac{\partial \Phi}{\partial \theta} = \frac{\partial \Psi}{\partial r} \qquad (4.2b)$$

Equations satisfying the Laplace equation are called harmonic. Equipotential lines (constant Φ) are orthogonal to streamlines (constant Ψ). The sum of several harmonic functions also satisfies the Laplace equation.

4.1   Irrotational flow around a circular cylinder

This section describes the combination of fundamental two-dimensional flow nets that define the flow configuration and the lift and drag forces around a circular cylinder.

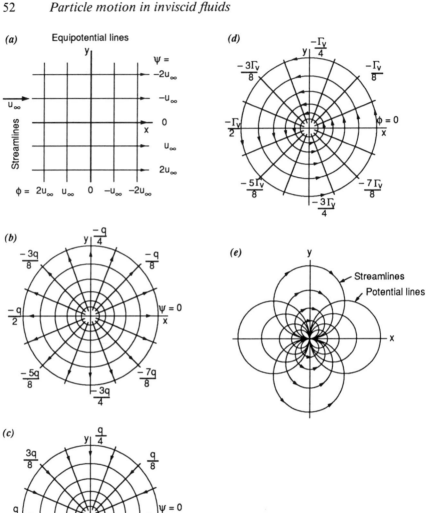

Figure 4.2. Two-dimensional flow nets: (a) rectilinear flow, (b) source, (c) sink, (d) vortex, (e) doublet

Rectilinear flow (Fig. 4.2a). Uniform flow velocity u_∞ along the x direction with origin at $(0, 0)$:

$$\Phi_1 = -u_\infty x = -u_\infty r \cos \theta \tag{4.3a}$$

$$\Psi_1 = -u_\infty y = -u_\infty r \sin \theta \tag{4.3b}$$

Source (Fig. 4.2b). The strength of a line source is equal to the volumetric flow rate $q = 2\pi r v_r$. Thus,

$$\Phi_{so} = -\frac{q}{4\pi} \ln(x^2 + y^2) = -\frac{q}{2\pi} \ln r \tag{4.4a}$$

$$\Psi_{so} = -\frac{q}{2\pi} \tan^{-1} \frac{y}{x} = -\frac{q\theta}{2\pi} \tag{4.4b}$$

Sink (Fig. 4.2c). A sink is a negative source; hence,

$$\Phi_{si} = \frac{q}{4\pi} \ln(x^2 + y^2) = \frac{q}{2\pi} \ln r \tag{4.5a}$$

$$\Psi_{si} = \frac{q}{2\pi} \tan^{-1} \frac{y}{x} = \frac{q\theta}{2\pi} \tag{4.5b}$$

Free vortex (Fig. 4.2d). The positive counterclockwise circulation $\Gamma_v = 2\pi r v_\theta$ of a free vortex with its center at the origin is a constant defining its strength:

$$\Phi_v = -\frac{\Gamma_v}{2\pi} \tan^{-1} \frac{y}{x} = -\frac{\Gamma_v \theta}{2\pi} \tag{4.6a}$$

$$\Psi_v = \frac{\Gamma_v}{4\pi} \ln(x^2 + y^2) = \frac{\Gamma_v}{2\pi} \ln r \tag{4.6b}$$

Doublet (Fig. 4.2e). A doublet, or dipole, is obtained when a source and a sink of equal strength are brought together in such a way that the product of their strength and the distance separating them remains constant:

$$\Phi_d = \frac{\Gamma_d}{2\pi} \left(\frac{x}{x^2 + y^2} \right) = \frac{\Gamma_d \cos \theta}{2\pi r} \tag{4.7a}$$

$$\Psi_d = -\frac{\Gamma_d}{2\pi} \left(\frac{y}{x^2 + y^2} \right) = -\frac{\Gamma_d \sin \theta}{2\pi r} \tag{4.7b}$$

These fundamental flow nets are combined to define the flow field around a two-dimensional circular cylinder, such as the near-surface flow field around a vertical bridge pier.

4.1.1   Flow field around a cylinder

The flow net with circulation past a circular cylinder of radius R is obtained by combining a rectilinear flow with a doublet of constant strength, $\Gamma_d = -2\pi u_\infty R^2$, and a counterclockwise free vortex of variable strength, Γ_v. The streamlines for flow around a cylinder without circulation ($\Gamma_v = 0$) are shown in Figure 4.3a; the effect of clockwise circulation ($\Gamma_v < 0$) is shown in Figure 4.3b.

$$\Phi_{cyl} = -u_\infty\left(x + \frac{R^2 x}{x^2 + y^2}\right) - \frac{\Gamma_v}{2\pi}\tan^{-1}\frac{y}{x}$$

$$= -u_\infty\left(r + \frac{R^2}{r}\right)\cos\theta - \frac{\Gamma_v \theta}{2\pi} \tag{4.8a}$$

$$\Psi_{cyl} = -u_\infty\left(y - \frac{R^2 y}{x^2 + y^2}\right) + \frac{\Gamma_v}{4\pi}\ln(x^2 + y^2)$$

$$= -u_\infty\left(r - \frac{R^2}{r}\right)\sin\theta + \frac{\Gamma_v}{2\pi}\ln r \tag{4.8b}$$

The velocity components around the cylinder are obtained after applying the operator in Equation (4.2) to the flow net from Equation (4.8):

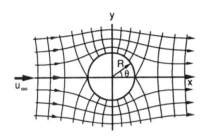

$$v_r = u_\infty\left(1 - \frac{R^2}{r^2}\right)\cos\theta \tag{4.9a}$$

$$v_\theta = -u_\infty\left(1 + \frac{R^2}{r^2}\right)\sin\theta$$

$$+ \frac{\Gamma_v}{2\pi r} \tag{4.9b}$$

(a) Without circulation, $\Gamma_v = 0$

where $x = r\cos\theta$, $y = r\sin\theta$, and v_θ is positive counterclockwise. At the cylinder surface, the radial velocity component $v_r = 0$, and v_θ vanishes only at the stagnation points.

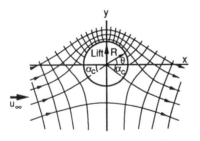

(b) With circulation, $\sin\alpha_c = \dfrac{\Gamma_v}{4\pi u_\infty R}$

Figure 4.3. Flow net around a cylinder

4.1.2   Lift and drag on a circular cylinder

The velocity at any point along the surface of the cylinder results from

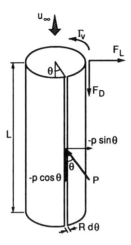

Figure 4.4. Lift and drag on a vertical cylinder

Equation (4.9b) at $r = R$. The surface pressure is then calculated from the velocity and the Bernoulli equation [Eq. (3.21b)], given the reference pressure $p = p_\infty = 0$, where $v = u_\infty$:

$$p = \frac{\rho_m}{2}\left[u_\infty^2 - \left(-2u_\infty \sin\theta + \frac{\Gamma_v}{2\pi R}\right)^2\right] \tag{4.10}$$

With reference to Figure 4.4, the drag force F_D per unit length is defined as the net fluid force in the direction parallel to the horizontal approach velocity u_∞ (toward the viewer), while the lift force F_L per unit length L is the net force in the normal direction (positive when $\theta = \pi/2$):

$$F_L = \int_0^{2\pi} -p \sin\theta\, LR\, d\theta$$

$$= \frac{\rho_m LR}{2} \int_0^{2\pi} -\left[u_\infty^2 - \left(-2u_\infty \sin\theta + \frac{\Gamma_v}{2\pi R}\right)^2\right]\sin\theta\, d\theta$$

$$= -\rho_m L u_\infty \Gamma_v \tag{4.11a}$$

$$F_D = \int_0^{2\pi} -p \cos\theta\, LR\, d\theta$$

$$= \frac{\rho_m LR}{2} \int_0^{2\pi} -\left[u_\infty^2 - \left(-2u_\infty \sin\theta + \frac{\Gamma_v}{2\pi R}\right)^2\right]\cos\theta\, d\theta = 0 \tag{4.11b}$$

It is concluded from this potential flow analysis for an inviscid fluid around a circular cylinder that circulation is necessary for the occurrence of lift forces. Drag forces cannot be generated in irrotational flows. Example 4.1 provides detailed calculations of lift and drag forces on a cylindrical surface.

Example 4.1   Application to lift and drag forces on a half-cylinder. Calculate the lift force per unit mass on the outer surface of an infinitely long $(L \to \infty)$ upper half-cylinder of radius R, assuming potential flow without circulation $(\Gamma_v = 0)$ (Fig. E4.1.1).

Step 1. The lift force $F_{L\frac{1}{2}}$ on the half-cylinder is

$$F_{L\frac{1}{2}} = \int_0^\pi (-p \sin \theta) LR \, d\theta$$

Substituting the pressure from Equation (4.10) without circulation $(\Gamma_v = 0)$,

$$F_{L\frac{1}{2}} = \int_0^\pi \left(-\frac{\rho_m LR}{2}\right)(u_\infty^2 - 4u_\infty^2 \sin^2 \theta) \sin \theta \, d\theta$$

$$F_{L\frac{1}{2}} = \frac{10}{6}\rho_m LRu_\infty^2 \tag{E4.1.1}$$

Step 2. The drag force $F_{D\frac{1}{2}}$ on the half-cylinder is

$$F_{D\frac{1}{2}} = \int_0^\pi (-p \cos \theta) LR \, d\theta = 0 \tag{E4.1.2}$$

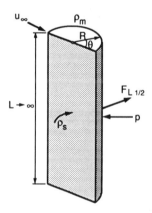

Figure E4.1.1. Lift and drag on a half-cylinder

Step 3. Assuming a solid half-cylinder of radius R and mass density $\rho_s = G\rho$, the acceleration a from the lift force per solid unit mass is

$$a = \frac{F_{L\frac{1}{2}}}{\text{mass}}$$

$$= \frac{F_{L\frac{1}{2}}}{\rho_s \forall} = \frac{10}{6}\frac{2\rho_m LRu_\infty^2}{\rho_s \pi R^2 L}$$

$$\simeq \frac{u_\infty^2}{GR} \tag{E4.1.3}$$

At a given velocity, the acceleration a is thus inversely proportional to the cylinder radius.

4.2   Irrotational flow around a sphere

This section describes fundamental three-dimensional flow patterns that can be combined to define the flow configuration around three-dimensional objects like a sphere. The discussion pertains to axisymmetric patterns for which $v_\varphi = 0$.

Three-dimensional rectilinear flow. Uniform flow velocity u_∞ along $x = r \cos \theta$ in spherical coordinates (r, θ, φ):

$$\Phi_{13} = -r u_\infty \cos \theta \tag{4.12a}$$

$$\Psi_{13} = -\frac{r^2 u_\infty}{2} \sin^2 \theta \tag{4.12b}$$

Three-dimensional source and sink. The strength of a point source is equal to the volumetric flow rate $Q = 4\pi r^2 v_r$. The flow net is given by

$$\Phi_{s3} = \frac{Q}{4\pi r} \tag{4.13a}$$

$$\Psi_{s3} = \frac{Q}{4\pi} \cos \theta \tag{4.13b}$$

A three-dimensional sink is a negative source for which the flow net is obtained after considering a negative discharge $-Q$ in Equations (4.13a) and (4.13b).

Three-dimensional doublet. As in the two-dimensional case, the doublet is obtained when a source and sink of equal strength are brought together in such a way that the product of their strength and the distance separating them remains constant. The resulting flow net is described by

$$\Phi_{d3} = -\frac{\Gamma_{d3} \cos \theta}{r^2} \tag{4.14a}$$

$$\Psi_{d3} = \frac{\Gamma_{d3} \sin^2 \theta}{r} \tag{4.14b}$$

4.2.1   Flow field around a sphere

For steady potential flow around a sphere of radius R in relative motion u_∞, the potential function Φ and the stream function Ψ in spherical coordinates, at a point (r, θ), are given from inserting a doublet of strength $\Gamma_{d3} = u_\infty R^3/2$ in three-dimensional rectilinear flow. The resulting flow net from Equations (4.12a,b) and (4.14a,b) is

$$\Phi = \frac{-u_\infty R^3}{2r^2} \cos\theta - u_\infty r \cos\theta \qquad (4.15a)$$

$$\Psi = \frac{u_\infty R^3}{2r} \sin^2\theta - u_\infty \frac{r^2}{2} \sin^2\theta \qquad (4.15b)$$

In three dimensions, the velocity components v_r and v_θ ($v_\varphi = 0$), in spherical coordinates, are given by the derivatives of the stream function Ψ and the potential function Φ from

$$v_r = \frac{-1}{r^2 \sin\theta} \frac{\partial \Psi}{\partial \theta} = -\frac{\partial \Phi}{\partial r} \qquad (4.16a)$$

$$v_\theta = \frac{1}{r \sin\theta} \frac{\partial \Psi}{\partial r} = -\frac{1}{r} \frac{\partial \Phi}{\partial \theta} \qquad (4.16b)$$

The velocity field is described by

$$v_r = u_\infty \cos\theta - u_\infty \frac{R^3}{r^3} \cos\theta \qquad (4.17a)$$

$$v_\theta = -u_\infty \sin\theta - \frac{u_\infty}{2} \frac{R^3}{r^3} \sin\theta \qquad (4.17b)$$

The velocity v at any point on the surface of the sphere ($r = R$) is obtained from $v = v_\theta$ because $v_r = 0$:

$$v = -\tfrac{3}{2} u_\infty \sin\theta \qquad (4.18)$$

4.2.2   Lift and drag forces on a sphere

The pressure distribution over the surface of the sphere is calculated from the Bernoulli equation and the velocity distribution [Eq. (4.18)]. Integrating the hydrostatic pressure distribution results in the buoyancy force obtained in Example 3.3. Considering the relative pressure $p - p_\infty$, the hydrodynamic pressure distribution associated with the velocity distribution around the sphere is given by

$$p - p_\infty = \frac{\rho_m u_\infty^2}{2} \left(1 - \frac{9}{4} \sin^2\theta\right) \qquad (4.19)$$

With reference to Figure 4.5, the lift force F_L is calculated from the pressure in the direction perpendicular to the main flow direction,

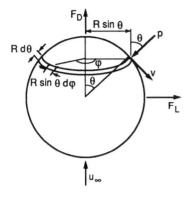

Figure 4.5. Lift and drag on a sphere

$-p \sin\theta \cos\varphi$, multiplied by the elementary surface area $R\,d\theta\,R \sin\theta\,d\varphi$ and integrated over the entire sphere, $0 < \varphi < 2\pi$, and $0 < \theta < \pi$:

$$F_L = \int_0^\pi \int_0^{2\pi} -p \sin\theta \cos\varphi\, R^2 \sin\theta\, d\varphi\, d\theta = 0 \tag{4.20a}$$

The drag force F_D is calculated from the integral of the pressure component in the main flow direction $-p \cos\theta$:

$$F_D = \int_0^\pi \int_0^{2\pi} -p \cos\theta\, R^2 \sin\theta\, d\varphi\, d\theta = 0 \tag{4.20b}$$

Irrotational flow around a sphere generates neither lift nor drag forces. Irrotational flow around asymmetric surfaces, however, generates lift and drag forces. An instructive application is presented in Example 4.2. The analysis of lift forces leads to the concept of particle equilibrium shown in Example 4.3.

Example 4.2   Application to lift and drag forces on a half-sphere. Calculate the drag force and the lift force from the pressure distribution of irrotational flow without circulation around a half-sphere (Fig. E4.2.1).

Drag force. The pressure component $-p \cos\theta$ in the flow direction is multiplied by the surface area $R\,d\theta\,R \sin\theta\,d\varphi$ and integrated over the entire surface of the half-sphere, $-\pi/2 < \varphi < \pi/2$ and $0 < \theta < \pi$:

$$F_D = \int_A -p \cos\theta\, dA$$

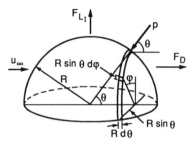

Figure E4.2.1. Lift and drag on a half-sphere

$$F_D = -\int_{-\pi/2}^{\pi/2} \int_0^{\pi} \frac{\rho_m u_\infty^2}{2}\left(1 - \frac{9}{4}\sin^2\theta\right)\cos\theta\, R\, d\theta\, R\sin\theta\, d\varphi \qquad (E4.2.1)$$

$$F_D = \pi R^2 \rho_m u_\infty^2 [0] = 0$$

Lift force. The lift force on a whole sphere is zero because of the symmetry. It is, however, instructive to calculate the lift force on a half-sphere (Fig. E4.2.1). The pressure component in the direction perpendicular to the main flow direction, $-p\sin\theta\cos\varphi$, is multiplied by the elementary surface area $dA = R\, d\theta\, R\sin\theta\, d\varphi$ and integrated over the half-sphere, $-\pi/2 < \varphi < \pi/2$ and $0 < \theta < \pi$:

$$F_{L1} = \int_A -p\sin\theta\cos\varphi\, dA$$

$$\qquad (E4.2.2)$$

$$F_{L1} = \int_0^{\pi} \int_{-\pi/2}^{\pi/2} -p\sin\theta\cos\varphi\, R\sin\theta\, d\varphi\, R\, d\theta = \frac{11\pi}{32}\rho_m u_\infty^2 R^2$$

When considering a solid upper half-sphere of mass density $\rho_s = \rho G$ and volume $\Psi = 4\pi R^3/6$, the acceleration a can be determined from the lift force per unit mass:

$$a = \frac{F_{L1}}{\rho_s \Psi} = \frac{11\pi\rho_m}{32\rho_s}\frac{u_\infty^2 R^2}{4\pi R^3}6 = \frac{33\rho_m}{64\rho_s}\frac{u_\infty^2}{R} \simeq 0.5\frac{u_\infty^2}{GR} \qquad (E4.2.3)$$

This expression for the acceleration, or lift force per unit mass, resembles that for flow around a cylindrical section [Eq. (E4.1.3)]. The coefficient depends on the shape of the particle. The most important conclusion from this analysis of particles in relative motion in inviscid fluids is that, for a given relative velocity u_∞, the lift force per unit volume (or mass) on curved surfaces is inversely proportional to the size of the particle. Thus, at large particle Reynolds numbers, fine particles are subjected to larger hydrodynamic accelerations than are coarse particles.

Example 4.3 Analysis of equilibrium of a sphere. Consider the hydrodynamic forces exerted on the upper half of the sphere illustrated in Figure E4.3.1 and determine the critical approach velocity u_c at which the sphere will be moved out of the pocket. The submerged weight F_S is given by subtracting the buoyancy force [Eq. (E3.3.1)] from the particle weight

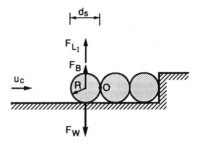

Figure E4.3.1. Equilibrium of a sphere

$$F_S = F_W - F_B = (\gamma_s - \gamma_m)\tfrac{4}{3}\pi R^3$$

The lift force F_{L1} on the upper half-sphere from Equation (E4.2.2) is

$$F_{L1} = \frac{11\pi}{32}\rho_m u_\infty^2 R^2$$

Equilibrium is obtained when $F_W - F_B - F_{L1} = 0$, which corresponds to $u_\infty = u_c$:

$$(\gamma_s - \gamma_m)\frac{4\pi}{3}R^3 = \frac{11\pi}{32}\rho_m u_c^2 R^2$$

$$u_c = \sqrt{(\gamma_s - \gamma_m)\frac{4}{3}R\frac{32}{11\rho_m}} \qquad\qquad\text{(E4.3.1)}$$

$$u_c \cong 1.4\sqrt{(G-1)gd_s}$$

Exercise

**4.1 Substitute the relationship of pressure p from Equation (4.19) into Equation (E4.2.2) and solve for F_{L1}.

*Problem 4.1

Determine the equation of pressure around a vertical half-cylinder from the Bernoulli equation [Eq. (3.21)] in a horizontal plane assuming $p_r = p_\infty = 0$ at $r = \infty$, where $v_r = u_\infty$.

Answer

$$p = \frac{\rho_m}{2}(u_\infty^2 - v^2) + p_\infty$$

*Problem 4.2

Calculate the lift force in lb on a 4-m-diameter semispherical tent under a 100-km/h wind.

**Problem 4.3

Plot and compare the distribution of surface velocity, pressure, and boundary shear stress for irrotational flow without circulation around a cylinder and a sphere of radius R.

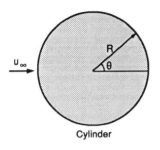
Cylinder

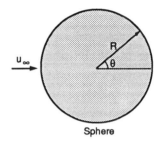
Sphere

Answer

$v = 0$ at $\theta = 0°$ and $180°$

$v_{\max}$ at $\theta = 90°$ and $270°$

$v_{\max}(\text{cylinder}) = 2u_\infty$

$v_{\max}(\text{sphere}) = 1.5u_\infty$

$p_{\max}$ at $\theta = 0°$ and $180°$

$p_{\min}$ at $\theta = 90°$ and $270°$

$p_{\max} = \rho u_\infty^2 / 2$

$\tau = 0$ everywhere

*Problem 4.4

Calculate the drag force on the outside surface of

(a) a quarter-spherical shell (*hint:* neglect the pressure inside the shell)

$$F_D = \int_{-\pi/2}^{\pi/2} \int_{\pi/2}^{\pi} -p \cos \theta \, R \, d\theta \, R \sin \theta \, d\varphi$$

(b) a half-spherical shell (*hint:* neglect the hydrostatic pressure distribution).

(a)

(b)

5

Particle motion in Newtonian fluids

At low rates of deformation and low sediment concentrations, sediment-laden flows obey Newton's law of deformation. The governing equations of motion are called the Navier–Stokes equations (Section 5.1), which are applied around a sphere to determine the flow field (Section 5.2), the drag force (Section 5.3), the fall velocity (Section 5.4), and the rate of energy dissipation (Section 5.5).

Viscosity is a fluid property that differentiates real fluids from ideal fluids. The viscosity of a fluid is a measure of resistance to flow. The dynamic viscosity of a Newtonian mixture μ_m is defined as the ratio of shear stress to the rate of deformation; thus, its dimensions are mass per unit length and time. In a Newtonian fluid the shear stress τ_{yx} acting in the x direction is proportional to the gradient in the y direction of the velocity component v_x:

$$\tau_{yx} = \mu_m \frac{dv_x}{dy} \tag{5.1}$$

The kinematic viscosity of a mixture ν_m is defined as the ratio of dynamic viscosity to mass density of the mixture ($\nu_m = \mu_m/\rho_m$), and the dimensions L^2/T do not involve mass dimensions.

5.1   Navier–Stokes equations

Shear stresses in a Newtonian fluid equal the product of dynamic viscosity of the mixture μ_m and rate of angular deformation; thus,

$$\tau_{yx} = \tau_{xy} = \mu_m \left(\frac{\partial v_x}{\partial y} + \frac{\partial v_y}{\partial x} \right) = \mu_m \oslash_z \tag{5.2a}$$

$$\tau_{zy} = \tau_{yz} = \mu_m \left(\frac{\partial v_y}{\partial z} + \frac{\partial v_z}{\partial y} \right) = \mu_m \oslash_x \tag{5.2b}$$

and

64

Table 5.1. *Navier–Stokes equations in Cartesian coordinates* (x, y, z)

---

x component

$$\frac{\partial v_x}{\partial t} + v_x\frac{\partial v_x}{\partial x} + v_y\frac{\partial v_x}{\partial y} + v_z\frac{\partial v_x}{\partial z} = g_x - \frac{1}{\rho_m}\frac{\partial p}{\partial x} + \frac{\mu_m}{\rho_m}\left(\frac{\partial^2 v_x}{\partial x^2} + \frac{\partial^2 v_x}{\partial y^2} + \frac{\partial^2 v_x}{\partial z^2}\right)$$

y component

$$\frac{\partial v_y}{\partial t} + v_x\frac{\partial v_y}{\partial x} + v_y\frac{\partial v_y}{\partial y} + v_z\frac{\partial v_y}{\partial z} = g_y - \frac{1}{\rho_m}\frac{\partial p}{\partial y} + \frac{\mu_m}{\rho_m}\left(\frac{\partial^2 v_y}{\partial x^2} + \frac{\partial^2 v_y}{\partial y^2} + \frac{\partial^2 v_y}{\partial z^2}\right)$$

z component

$$\frac{\partial v_z}{\partial t} + v_x\frac{\partial v_z}{\partial x} + v_y\frac{\partial v_z}{\partial y} + v_z\frac{\partial v_z}{\partial z} = g_z - \frac{1}{\rho_m}\frac{\partial p}{\partial z} + \frac{\mu_m}{\rho_m}\left(\frac{\partial^2 v_z}{\partial x^2} + \frac{\partial^2 v_z}{\partial y^2} + \frac{\partial^2 v_z}{\partial z^2}\right)$$

$$\tau_{xz} = \tau_{zx} = \mu_m\left(\frac{\partial v_z}{\partial x} + \frac{\partial v_x}{\partial z}\right) = \mu_m \oslash_y \tag{5.2c}$$

The normal stresses of an isotropic Newtonian fluid are also related to pressure p, viscosity μ_m, and velocity gradients. The relationships for the normal stresses σ_x, σ_y, and σ_z are

$$\sigma_x = -p + 2\mu_m\frac{\partial v_x}{\partial x} - \frac{2\mu_m}{3}\left(\frac{\partial v_x}{\partial x} + \frac{\partial v_y}{\partial y} + \frac{\partial v_z}{\partial z}\right) \tag{5.3a}$$

$$\sigma_y = -p + 2\mu_m\frac{\partial v_y}{\partial y} - \frac{2\mu_m}{3}\left(\frac{\partial v_x}{\partial x} + \frac{\partial v_y}{\partial y} + \frac{\partial v_z}{\partial z}\right) \tag{5.3b}$$

$$\sigma_z = -p + 2\mu_m\frac{\partial v_z}{\partial z} - \frac{2\mu_m}{3}\left(\frac{\partial v_x}{\partial x} + \frac{\partial v_y}{\partial y} + \frac{\partial v_z}{\partial z}\right) \tag{5.3c}$$

For incompressible mixtures, the terms in parentheses in Equations (5.3) can be dropped because they correspond to the continuity equation [Eq. (3.10)].

Substitution of these stress tensor relationships [Eqs. (5.2) and (5.3)] into the equations of motion [Eqs. (3.12)] gives the complete set of equations of motion shown in Tables 5.1–5.3 for Newtonian mixtures in the laminar regime (low Reynolds numbers). These equations were developed by Navier, Cauchy, Poisson, Saint-Venant, and Stokes. They are commonly referred to as the Navier–Stokes equations.

5.2   Newtonian flow around a sphere

The vorticity components $\otimes_x$, $\otimes_y$, and $\otimes_z$ defined previously [Eq. (3.4)] have been shown to satisfy Equation (3.6), which can be used to rewrite

Table 5.2. *Stress tensor and Navier–Stokes equations in cylindrical coordinates* (r, θ, z)

$$\sigma_{rr} = -p + \mu_{\mathrm{m}}\left[2\frac{\partial v_r}{\partial r} - \frac{2}{3}(\nabla \cdot v)\right]$$

$$\sigma_{\theta\theta} = -p + \mu_{\mathrm{m}}\left[2\left(\frac{1}{r}\frac{\partial v_\theta}{\partial \theta} + \frac{v_r}{r}\right) - \frac{2}{3}(\nabla \cdot v)\right]$$

$$\sigma_{zz} = -p + \mu_{\mathrm{m}}\left[2\frac{\partial v_z}{\partial z} - \frac{2}{3}(\nabla \cdot v)\right]$$

$$\tau_{r\theta} = \tau_{\theta r} = \mu_{\mathrm{m}}\left[r\frac{\partial}{\partial r}\left(\frac{v_\theta}{r}\right) + \frac{1}{r}\frac{\partial v_r}{\partial \theta}\right]$$

$$\tau_{\theta z} = \tau_{z\theta} = \mu_{\mathrm{m}}\left[\frac{\partial v_\theta}{\partial z} + \frac{1}{r}\frac{\partial v_z}{\partial \theta}\right]$$

$$\tau_{zr} = \tau_{rz} = \mu_{\mathrm{m}}\left[\frac{\partial v_z}{\partial r} + \frac{\partial v_r}{\partial z}\right]$$

where

$$(\nabla \cdot v) = \frac{1}{r}\frac{\partial}{\partial r}(rv_r) + \frac{1}{r}\frac{\partial v_\theta}{\partial \theta} + \frac{\partial v_z}{\partial z}$$

r component

$$\frac{\partial v_r}{\partial t} + v_r\frac{\partial v_r}{\partial r} + \frac{v_\theta}{r}\frac{\partial v_r}{\partial \theta} - \frac{v_\theta^2}{r} + v_z\frac{\partial v_r}{\partial z}$$

$$= g_r - \frac{1}{\rho_{\mathrm{m}}}\frac{\partial p}{\partial r} + \frac{\mu_{\mathrm{m}}}{\rho_{\mathrm{m}}}\left[\frac{\partial}{\partial r}\left(\frac{1}{r}\frac{\partial}{\partial r}(rv_r)\right) + \frac{1}{r^2}\frac{\partial^2 v_r}{\partial \theta^2} - \frac{2}{r^2}\frac{\partial v_\theta}{\partial \theta} + \frac{\partial^2 v_r}{\partial z^2}\right]$$

θ component

$$\frac{\partial v_\theta}{\partial t} + v_r\frac{\partial v_\theta}{\partial r} + \frac{v_\theta}{r}\frac{\partial v_\theta}{\partial \theta} + \frac{v_r v_\theta}{r} + v_z\frac{\partial v_\theta}{\partial z}$$

$$= g_\theta - \frac{1}{r\rho_{\mathrm{m}}}\frac{\partial p}{\partial \theta} + \frac{\mu_{\mathrm{m}}}{\rho_{\mathrm{m}}}\left[\frac{\partial}{\partial r}\left(\frac{1}{r}\frac{\partial}{\partial r}(rv_\theta)\right) + \frac{1}{r^2}\frac{\partial^2 v_\theta}{\partial \theta^2} + \frac{2}{r^2}\frac{\partial v_r}{\partial \theta} + \frac{\partial^2 v_\theta}{\partial z^2}\right]$$

z component

$$\frac{\partial v_z}{\partial t} + v_r\frac{\partial v_z}{\partial r} + \frac{v_\theta}{r}\frac{\partial v_z}{\partial \theta} + v_z\frac{\partial v_z}{\partial z}$$

$$= g_z - \frac{1}{\rho_{\mathrm{m}}}\frac{\partial p}{\partial z} + \frac{\mu_{\mathrm{m}}}{\rho_{\mathrm{m}}}\left[\frac{1}{r}\frac{\partial}{\partial r}\left(\frac{r\partial v_z}{\partial r}\right) + \frac{1}{r^2}\frac{\partial^2 v_z}{\partial \theta^2} + \frac{\partial^2 v_z}{\partial z^2}\right]$$

Table 5.3. *Stress tensor and Navier–Stokes equations in*
spherical coordinates (r, θ, φ)

$$\sigma_{rr} = -p + \mu_m \left[2\frac{\partial v_r}{\partial r} - \frac{2}{3}(\nabla \cdot v) \right]$$

$$\sigma_{\theta\theta} = -p + \mu_m \left[2\left(\frac{1}{r}\frac{\partial v_\theta}{\partial \theta} + \frac{v_r}{r} \right) - \frac{2}{3}(\nabla \cdot v) \right]$$

$$\sigma_{\varphi\varphi} = -p + \mu_m \left[2\left(\frac{1}{r \sin\theta}\frac{\partial v_\varphi}{\partial \varphi} + \frac{v_r}{r} + \frac{v_\theta \cot\theta}{r} \right) - \frac{2}{3}(\nabla \cdot v) \right]$$

$$\tau_{r\theta} = \tau_{\theta r} = \mu_m \left[r\frac{\partial}{\partial r}\left(\frac{v_\theta}{r} \right) + \frac{1}{r}\frac{\partial v_r}{\partial \theta} \right]$$

$$\tau_{\theta\varphi} = \tau_{\varphi\theta} = \mu_m \left[\frac{\sin\theta}{r}\frac{\partial}{\partial \theta}\left(\frac{v_\varphi}{\sin\theta} \right) + \frac{1}{r \sin\theta}\frac{\partial v_\theta}{\partial \varphi} \right]$$

$$\tau_{\varphi r} = \tau_{r\varphi} = \mu_m \left[\frac{1}{r \sin\theta}\frac{\partial v_r}{\partial \varphi} + r\frac{\partial}{\partial r}\left(\frac{v_\varphi}{r} \right) \right]$$

$$(\nabla \cdot v) = \frac{1}{r^2}\frac{\partial}{\partial r}(r^2 v_r) + \frac{1}{r \sin\theta}\frac{\partial}{\partial \theta}(v_\theta \sin\theta) + \frac{1}{r \sin\theta}\frac{\partial v_\varphi}{\partial \varphi}$$

$$\nabla^2 = \frac{1}{r^2}\frac{\partial}{\partial r}\left(r^2\frac{\partial}{\partial r} \right) + \frac{1}{r^2 \sin\theta}\frac{\partial}{\partial \theta}\left(\sin\theta\frac{\partial}{\partial \theta} \right) + \frac{1}{r^2 \sin^2\theta}\left(\frac{\partial^2}{\partial \varphi^2} \right)$$

r component

$$\frac{\partial v_r}{\partial t} + v_r\frac{\partial v_r}{\partial r} + \frac{v_\theta}{r}\frac{\partial v_r}{\partial \theta} + \frac{v_\varphi}{r \sin\theta}\frac{\partial v_r}{\partial \varphi} - \frac{v_\theta^2 + v_\varphi^2}{r}$$

$$= g_r - \frac{1}{\rho_m}\frac{\partial p}{\partial r} + \frac{\mu_m}{\rho_m}\left[\nabla^2 v_r - \frac{2}{r^2}v_r - \frac{2}{r^2}\frac{\partial v_\theta}{\partial \theta} - \frac{2}{r^2}v_\theta \cot\theta - \frac{2}{r^2 \sin\theta}\frac{\partial v_\varphi}{\partial \varphi} \right]$$

θ component

$$\frac{\partial v_\theta}{\partial t} + v_r\frac{\partial v_\theta}{\partial r} + \frac{v_\theta}{r}\frac{\partial v_\theta}{\partial \theta} + \frac{v_\varphi}{r \sin\theta}\frac{\partial v_\theta}{\partial \varphi} + \frac{v_r v_\theta}{r} - \frac{v_\varphi^2 \cot\theta}{r}$$

$$= g_\theta - \frac{1}{\rho_m r}\frac{\partial p}{\partial \theta} + \frac{\mu_m}{\rho_m}\left[\nabla^2 v_\theta + \frac{2}{r^2}\frac{\partial v_r}{\partial \theta} - \frac{v_\theta}{r^2 \sin^2\theta} - \frac{2 \cos\theta}{r^2 \sin^2\theta}\frac{\partial v_\varphi}{\partial \varphi} \right]$$

φ component

$$\frac{\partial v_\varphi}{\partial t} + v_r\frac{\partial v_\varphi}{\partial r} + \frac{v_\theta}{r}\frac{\partial v_\varphi}{\partial \theta} + \frac{v_\varphi}{r \sin\theta}\frac{\partial v_\varphi}{\partial \varphi} + \frac{v_\varphi v_r}{r} + \frac{v_\theta v_\varphi}{r}\cot\theta$$

$$= g_\varphi - \frac{1}{\rho_m r \sin\theta}\frac{\partial p}{\partial \varphi} + \frac{\mu_m}{\rho_m}\left[\nabla^2 v_\varphi - \frac{v_\varphi}{r^2 \sin^2\theta} + \frac{2}{r^2 \sin\theta}\frac{\partial v_r}{\partial \varphi} + \frac{2 \cos\theta}{r^2 \sin^2\theta}\frac{\partial v_\theta}{\partial \varphi} \right]$$

the Navier–Stokes equations in Table 5.1 for an incompressible fluid in a form similar to Equations (3.19):

$$\frac{\partial v_x}{\partial t} - v_y \otimes_z + v_z \otimes_y = -\frac{g}{\partial x}\frac{\partial H}{\partial x} + \nu_m \nabla^2 v_x \qquad (5.4a)$$

$$\frac{\partial v_y}{\partial t} - v_z \otimes_x + v_x \otimes_z = -\frac{g}{\partial y}\frac{\partial H}{\partial y} + \nu_m \nabla^2 v_y \qquad (5.4b)$$

$$\frac{\partial v_z}{\partial t} - v_x \otimes_y + v_y \otimes_x = -\frac{g}{\partial z}\frac{\partial H}{\partial z} + \nu_m \nabla^2 v_z \qquad (5.4c)$$

in which H represents the Bernoulli sum from Equation (3.20a).

After elimination of the Bernoulli sum H through cross-differentiation (see Exercise 5.3), the foregoing equations become

$$\frac{d\otimes_x}{dt} = \otimes_x \frac{\partial v_x}{\partial x} + \otimes_y \frac{\partial v_x}{\partial y} + \otimes_z \frac{\partial v_x}{\partial z} + \nu_m \nabla^2 \otimes_x \qquad (5.5a)$$

$$\frac{d\otimes_y}{dt} = \otimes_x \frac{\partial v_y}{\partial x} + \otimes_y \frac{\partial v_y}{\partial y} + \otimes_z \frac{\partial v_y}{\partial z} + \nu_m \nabla^2 \otimes_y \qquad (5.5b)$$

$$\frac{d\otimes_z}{dt} = \otimes_x \frac{\partial v_z}{\partial x} + \otimes_y \frac{\partial v_z}{\partial y} + \otimes_z \frac{\partial v_z}{\partial z} + \nu_m \nabla^2 \otimes_z \qquad (5.5c)$$

These are the equations governing the diffusion of vorticity.

Irrotational flow of a viscous fluid is possible because the conditions $\otimes_x = \otimes_y = \otimes_z = 0$ are also solutions to the Navier–Stokes equations from which the vorticity equations were derived. However, such flows are possible only when the solid boundary moves at the same velocity as the fluid at the boundary.

The equations of diffusion of vorticity are analogous to the law of conduction of heat. It is evident from this analogy that vorticity cannot originate from the interior of a viscous fluid, but must diffuse inward from the boundary.

The vorticity transport equations can also be written in terms of the stream function Ψ. From the definition of the two-dimensional stream function ($v_x = -\partial\Psi/\partial y$ and $v_y = \partial\Psi/\partial x$) the continuity equation is satisfied and the vorticity transport equation becomes

$$\frac{\partial \nabla^2 \Psi}{\partial t} - \frac{\partial \Psi}{\partial y}\frac{\partial \nabla^2 \Psi}{\partial x} + \frac{\partial \Psi}{\partial x}\frac{\partial \nabla^2 \Psi}{\partial y} = \nu_m \nabla^4 \Psi \qquad (5.6)$$

This fourth-order nonlinear partial differential equation contains only one unknown in Ψ. The inertial terms on the left-hand side are balanced

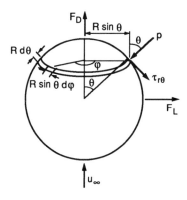

Figure 5.1. Creeping flow past a sphere

by the frictional terms on the right-hand side of the equation. The physical significance of the stream function is that lines of constant value of Ψ are streamlines.

When the viscous forces are considerably larger than the inertial forces, the left-hand side of Equation (5.6) vanishes and the vorticity transport equation reduces to the following linear equation:

$$\nabla^4 \Psi = 0 \qquad (5.7)$$

Creeping motion of a small sphere of radius R in a Newtonian fluid can be obtained by applying Equation (5.7) to the steady motion of a sphere at a small relative velocity u_∞ (Fig. 5.1).

The velocity distribution around a sphere of radius R in creeping motion is given by

$$\frac{v_r}{u_\infty} = \left[1 - \frac{3}{2}\left(\frac{R}{r}\right) + \frac{1}{2}\left(\frac{R}{r}\right)^3\right]\cos\theta \qquad (5.8a)$$

and

$$\frac{v_\theta}{u_\infty} = -\left[1 - \frac{3}{4}\left(\frac{R}{r}\right) - \frac{1}{4}\left(\frac{R}{r}\right)^3\right]\sin\theta \qquad (5.8b)$$

The corresponding pressure and shear stress distributions are found analytically (given here without derivation) from the stress tensor and the Navier–Stokes equations in spherical coordinates (Table 5.3). The pressure can be divided into the ambient pressure p_r, the hydrostatic pressure p_h, and the dynamic pressure p_d, such that $p = p_h + p_d$:

$$p_h = p_r - \rho_m g R \cos\theta \qquad (5.9a)$$

$$p_d = -\frac{3}{2}\frac{\mu_m u_\infty}{R}\left(\frac{R}{r}\right)^2 \cos\theta \qquad (5.9b)$$

and

$$\tau = \tau_{r\theta} = -\frac{3\mu_m u_\infty}{2R}\left(\frac{R}{r}\right)^4 \sin\theta \qquad (5.9c)$$

The quantity p_r is the ambient pressure far away from the sphere ($\theta = 90°$ or $270°$ and $R \to \infty$) and $-\rho_m g R \cos\theta$ is the hydrostatic pressure contribution due to the fluid mixture weight. Shear stress is positive in the direction of increasing angle θ.

5.3   Drag force on a sphere

Besides the buoyancy force resulting from integrating the hydrostatic pressure distribution on a sphere (Example 3.3), the drag force exerted by the moving fluid around the sphere is computed by integrating the stress tensor over the sphere surface. The drag force has two components: (1) The surface drag results from the integration of the shear stress distribution; and (2) the form drag arises from the integration of the hydrodynamic pressure distribution.

5.3.1   Surface drag

With reference to Figure 5.1, the shear stress $\tau = \tau_{r\theta}$ at each point on the spherical surface $r = R$ is the tangential force in the increasing θ direction per unit area of spherical surface. The shear stress component in the flow direction, $-\tau \sin \theta$, is multiplied by the elementary area, $R^2 \sin \theta \, d\theta \, d\varphi$, and integrated over the spherical surface to give the surface drag F_D':

$$F_D' = \int_0^{2\pi} \int_0^{\pi} -\tau \sin \theta \, R^2 \sin \theta \, d\theta \, d\varphi \qquad (5.10)$$

The shear stress distribution τ at the surface of the sphere from Equation (5.9c) is substituted into the integral in Equation (5.10) to give the surface drag force:

$$F_D' = 4\pi \mu_m R u_\infty \qquad (5.11) \; \blacklozenge$$

5.3.2   Form drag

At each point on the surface of the sphere, the dynamic pressure p_d acts perpendicularly to the surface, the vertical component of which is $-p_d \cos \theta$. We now multiply this local force per unit area by the surface area on which it is applied, $R^2 \sin \theta \, d\theta \, d\varphi$, and integrate over the surface of the sphere to get the resultant vertical force, called form drag force F_D'':

$$F_D'' = \int_0^{2\pi} \int_0^{\pi} -p_d \cos \theta \, R^2 \sin \theta \, d\theta \, d\varphi \qquad (5.12)$$

Substituting the dynamic pressure p_d from Equation (5.9b) into the integral gives the form drag force F_D'':

$$F_D'' = 2\pi\mu_m R u_\infty \tag{5.13} \blacklozenge$$

Hence, the total drag force F_D exerted by the motion of a Newtonian viscous fluid around the sphere is given by the sum of Equations (5.11) and (5.13):

$$F_D = F_D' + F_D'' = \underbrace{4\pi\mu_m R u_\infty}_{\text{surface drag}} + \underbrace{2\pi\mu_m R u_\infty}_{\text{form drag}} = \underbrace{6\pi\mu_m R u_\infty}_{\text{total drag}} \tag{5.14}$$

The buoyancy force F_B resulting from the hydrostatic pressure distribution from Example 3.3 is added to the drag force F_D from Equation (5.14) to give the total force F:

$$F = F_B + F_D = \underbrace{\tfrac{4}{3}\pi\gamma_m R^3}_{\text{buoyancy}} + \underbrace{6\pi\mu_m R u_\infty}_{\text{total drag}} \tag{5.15}$$

The total upward force F exerted on a sphere falling at a velocity u_∞ is thus the sum of the hydrostatic (buoyancy force) and hydrodynamic (total drag) components.

Example 5.1 illustrates the application of the surface drag and form drag exerted on a spherical particle.

Example 5.1 Analysis of particle equilibrium. Assume a uniformly distributed approach velocity u_∞ on the top half-sphere of radius R (Fig. E5.1.1). Determine from equilibrium conditions the critical velocity u_c at which the sphere will move out of the pocket. Equilibrium is governed by the sum of moments about point O. Consider the form drag and the surface drag exerted on the top half of the sphere. Notice that the lift force vanishes due to the symmetrical distribution of shear stress and pressure about the vertical axis.

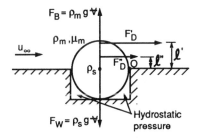

Figure E5.1.1. Particle equilibrium

Step 1. With reference to Figure 5.1, the form drag F_D'' on the upper half-spherical shell is given by

$$F_D'' = \int_0^\pi \int_0^\pi -p_d \cos\theta \, R^2 \sin\theta \, d\theta \, d\varphi$$

$$F_D'' = \int_0^\pi \int_0^\pi \frac{3}{2} \frac{\mu_m u_\infty}{R} \cos^2\theta \sin\theta \, R^2 \, d\theta \, d\varphi \qquad \text{[from Eq. (5.9b)]}$$

$$F_D'' = \pi \mu_m u_\infty R$$

The corresponding clockwise moment M_D'' about point O is

$$M_D'' = \int_0^\pi \int_0^\pi -p_d \cos\theta \, R^2 \sin\theta \, (R\sin\theta) \, d\theta \, d\varphi$$

$$M_D'' = \int_0^\pi \int_0^\pi \frac{3}{2} \frac{\mu_m u_\infty}{R} \cos^2\theta \, R^3 \sin^2\theta \, d\theta \, d\varphi$$

$$M_D'' = \frac{3\pi^2}{16} \mu_m u_\infty R^2$$

The moment arm $l'' = M_D''/F_D''$ is calculated from

$$l'' = \frac{M_D''}{F_D''} = \frac{3\pi^2}{16} \frac{\mu_m u_\infty R^2}{\pi \mu_m u_\infty R} = \frac{3\pi}{16} R$$

Step 2. The surface drag F_D', clockwise moment M_D', and moment arm l' on the upper spherical shell are, respectively:

$$F_D' = \int_0^\pi \int_0^\pi -\tau \sin\theta \, R^2 \sin\theta \, d\theta \, d\varphi$$

$$F_D' = \int_0^\pi \int_0^\pi \frac{3}{2} \mu_m \frac{u_\infty}{R} \sin\theta \sin\theta \, R^2 \sin\theta \, d\theta \, d\varphi \qquad \text{[from Eq. (5.9c)]}$$

$$F_D' = 2\pi \mu_m u_\infty R$$

$$M_D' = \int_0^\pi \int_0^\pi -\tau \sin\theta \, R^2 \sin\theta \, (R\sin\theta) \, d\theta \, d\varphi$$

$$M_D' = \int_0^\pi \int_0^\pi \frac{3\mu_m}{2} \frac{u_\infty}{R} \sin\theta \sin\theta \, R^2 \sin\theta \, R \sin\theta \, d\theta \, d\varphi$$

$$M_D' = \frac{9\pi^2}{16} \mu_m u_\infty R^2$$

$$l' = \frac{M_D'}{F_D'} = \frac{9}{16} \frac{\pi^2 \mu_m u_\infty R^2}{2\pi \mu_m u_\infty R} = \frac{9\pi R}{32}$$

Step 3. Equilibrium defined when $u_\infty = u_c$ occurs when the sum of moments about point O vanishes:

$$\Sigma M_0 = F_B R + M'_D + M''_D - F_W R = 0$$

$$(\gamma_s - \gamma_m)\frac{4}{3}\pi R^4 = \frac{9\pi^2}{16}\mu_m u_c R^2 + \frac{3\pi^2}{16}\mu_m u_c R^2 \qquad \text{(E5.1.1)}$$

$$u_c = \frac{16}{9\pi}\frac{(\gamma_s - \gamma_m)R^2}{\mu_m}$$

Further assuming that the critical shear stress $\tau_c \cong \mu_m u_c / R$, with $R = d_s/2$, one obtains

$$\frac{\tau_c}{(\gamma_s - \gamma_m)d_s} = \frac{2\mu_m u_c}{(\gamma_s - \gamma_m)d_s^2} = \frac{8}{9\pi} = 0.283 \qquad \text{(E5.1.2)}$$

This critical shear stress value approximately describes the beginning of motion of noncohesive fine particles (Chapter 7).

5.4   Drag coefficient and fall velocity

5.4.1   Drag coefficient

With reference to the dimensional analysis in Example 2.1 describing the flow around a particle of diameter d_s, it was inferred in Equation (E2.1.3) that the drag coefficient C_D could be written as a function of the particle Reynolds number $\mathrm{Re}_p = u_\infty d_s / \nu_m$. For creeping flow of a mixture (ρ_m, μ_m) around a sphere, this relationship is obtained after substituting F_D from Equation (5.14) into Equation (E2.1.2):

$$C_D = \frac{8F_D}{\rho_m \pi u_\infty^2 d_s^2} = \frac{24\nu_m}{u_\infty d_s} = \frac{24}{\mathrm{Re}_p} \qquad (5.16)$$

This equation is valid when $\mathrm{Re}_p < 0.1$. Oseen (1927) and Goldstein (1929) included the inertial terms on the left-hand side of the vorticity transport equations [Eq. (5.6)]. The drag coefficient when $\mathrm{Re}_p < 1$ is better approximated by

$$C_D = \frac{24}{\mathrm{Re}_p}\left(1 + \frac{3}{16}\mathrm{Re}_p - \frac{19}{1,280}\mathrm{Re}_p^2 + \frac{71}{20,480}\mathrm{Re}_p^3 + \cdots\right) \qquad (5.17)$$

Rubey (1933) followed with a simple approximation of the drag coefficient encompassing a wide range of particle Reynolds numbers:

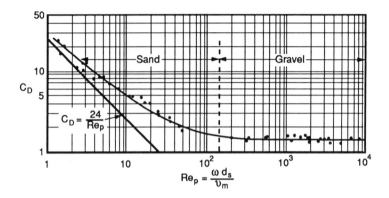

Figure 5.2. Drag coefficient for natural sands and gravels

$$C_D = \frac{24}{\text{Re}_p} + 2 \tag{5.18}$$

The experimental values of the drag coefficient of Engelund and Hansen (1967) shown in Figure 5.2 should be used for large values of the particle Reynolds number ($\text{Re}_p < 10^4$). The following relationship of the drag coefficient seems best suited to natural sands and gravels:

$$C_D = \frac{24}{\text{Re}_p} + 1.5 \tag{5.19}$$

5.4.2   Fall velocity

Starting from rest, a particle of density exceeding that of the surrounding fluid ($G > 1$) will accelerate in the downward direction until it reaches an equilibrium fall velocity ω. The equilibrium fall velocity ($\omega = u_\infty$) of a small solid sphere falling in creeping motion under its own weight ($F_W = \gamma_s \forall$) in a viscous fluid is calculated by substituting the particle weight F_W for the force F in Equation (5.15):

$$F_W = \frac{\pi}{6} d_s^3 \rho_s g = \frac{\pi}{6} d_s^3 \rho_m g + 3\pi \mu_m d_s \omega \tag{5.20}$$

Solving Equation (5.20) for the vall velocity ω as a function of the particle diameter d_s gives Equation (5.21a) in a mixture and Equation (5.21b) for the settling velocity ω_0 in clear water:

$$\omega = \frac{1}{18}\frac{\gamma_s - \gamma_m}{\mu_m}d_s^2 \quad \text{in a mixture where } \mathrm{Re_p} < 0.1 \qquad (5.21a) \blacklozenge$$

$$\omega_0 = \frac{1}{18}\frac{(G-1)g}{\nu}d_s^2 \quad \text{in clear water where } \mathrm{Re_p} < 0.1 \qquad (5.21b) \blacklozenge$$

This equation is valid for small particles ($d_s < 0.1$ mm in water) falling in viscous fluids ($\mathrm{Re_p} < 0.1$). In a more general form, the fall velocity is expressed as a function of the drag coefficient C_D from Equations (5.16) and (5.20) after replacing F_D by $F_W - F_B$ with $G = \gamma_s/\gamma$:

$$\omega = \left[\frac{4}{3}\frac{\gamma_s - \gamma_m}{\rho_m}\frac{d_s}{C_D}\right]^{1/2} \quad \text{in a mixture} \qquad (5.22a) \blacklozenge$$

$$\omega_0 = \left[\frac{4}{3}(G-1)\frac{gd_s}{C_D}\right]^{1/2} \quad \text{in clear water} \qquad (5.22b) \blacklozenge$$

Notice that the fall velocity of coarse particles ($C_D = 1.5$, or $d_s \geq 1$ mm in water) is roughly equal to $\omega_0 \cong \sqrt{(G-1)gd_s}$. It is instructive to compare this result with Equation (E4.3.1).

Rubey's approximate formulation of the fall velocity in clear water based on $C_D = (24\nu/\omega d_s) + 2$ is given by

$$\omega_0 = \frac{1}{d_s}\left[\sqrt{\frac{2g}{3}(G-1)d_s^3 + 36\nu^2} - 6\nu\right] \qquad (5.23a) \blacklozenge$$

or

$$\omega_0 = \left[\sqrt{\frac{2}{3} + \frac{36\nu^2}{(G-1)gd_s^3}} - \sqrt{\frac{36\nu^2}{(G-1)gd_s^3}}\right]\sqrt{(G-1)gd_s} \qquad (5.23b)$$

Except for its use in Einstein's bedload equation (Section 9.1.3) this formulation is rarely seen in practice.

A similar formula based on the drag coefficient of sand particles in Figure 5.2 with $C_D = (24\nu_m/\omega d_s) + 1.5$ gives

$$\omega = \frac{8\nu_m}{d_s}\{[1 + 0.0139d_*^3]^{0.5} - 1\} \qquad (5.23c) \blacklozenge$$

where the dimensionless particle diameter d_* is defined as

$$d_* = d_s\left[\frac{(G-1)g}{\nu_m^2}\right]^{1/3} \qquad (5.23d)$$

Equation (5.23c) estimates the fall velocity of particles under a wide range of Reynolds numbers. Approximate values of fall velocity in clear water are given in Table 5.4.

Table 5.4. *Clear water fall velocity ω_0 and dimensionless particle diameter d_**

Class name	Particle diameter (mm)	ω_0 near 0°C (mm/s)	ω_0 at 10°C (mm/s)	ω_0 at 20°C (mm/s)	d_* near 0°C	d_* at 10°C	d_* at 20°C
Boulder							
Very large	>2,048	5,430	5,430	5,430	35,140	43,271	51,807
Large	>1,024	3,839	3,839	3,839	17,570	21,635	25,903
Medium	>512	2,715	2,715	2,715	8,785	10,817	12,951
Small	>256	1,919	1,919	1,919	4,382	5,409	6,475
Cobble							
Large	>128	1,357	1,357	1,357	2,196	2,704	3,237
Small	>64	959	959	959	1,098	1,352	1,618
Gravel							
Very coarse	>32	678	678	678	549	676	809
Coarse	>16	479	479	479	274	338	404
Medium	>8	337	338	338	137	169	202
Fine	>4	236	237	238	68	84	101
Very fine	>2	162	164	167	34	42	50
Sand							
Very coarse	>1	106	109	112	17.1	21	25
Coarse	>0.5	61	66.4	70.3	8.58	10.5	12.6
Medium	>0.25	25.7	31.3	36	4.29	5.3	6.32
Fine	>0.125	7.6	10.1	12.8	2.14	2.6	3.16
Very fine	>0.0625	1.9	2.66	3.47	1.07	1.3	1.58

Silt							
Coarse	>0.031	0.49^a	0.67^a	0.88^a	0.536	0.66	0.79
Medium	>0.016	0.12^a	0.167^a	0.22^a	0.268	0.33	0.395
Fine	>0.008	0.031^a	0.042^a	0.055^a	0.134	0.165	0.197
Very fine	>0.004	0.0076^a	0.010^a	0.014^a	0.067	0.082	0.099
Clay							
Coarse	>0.002	1.9×10^{-3a}	2.6×10^{-3a}	3.4×10^{-3a}	0.033	0.041	0.049
Medium	>0.001	4.8×10^{-4a}	6.5×10^{-4a}	8.6×10^{-4a}	0.016	0.021	0.025
Fine	>0.0005	1.2×10^{-4a}	1.63×10^{-4a}	2.1×10^{-4a}	0.008	0.010	0.012
Very fine	>0.00024	2.9×10^{-5a}	4.1×10^{-5a}	5.3×10^{-5a}	0.004	0.005	0.006

[a] Possible flocculation (see Section 5.4.3).

5.4.3   *Flocculation*

Flocculation is the tendency of very fine sediments to aggregate and settle as a flocculated mass. In general, flocculation is enhanced at high sediment concentration, high salt content, and high fluid temperature. According to Migniot (1989), the settling velocity of flocculated particles ω_f can be calculated given the settling velocity of dispersed particles ω from

$$\omega_f = \frac{250}{d_s^2}\omega \quad \text{when} \quad d_s < 40\,\mu\text{m} \tag{5.24}$$

where d_s is the particle diameter in micrometers ($1\,\mu\text{m} = 10^{-6}$ m).

Considering Stokes' law, the flocculated settling velocity is approximately 0.15 to 0.6 mm/s and does not vary much with the particle size. This flocculated settling velocity is comparable to the settling of medium to coarse silt particles. Flocculation is not important on particles larger than coarse silts ($d_s > 0.04$ mm).

Dispersion can be secured in laboratory settling experiments by adding a 1% dilution of a dispersion agent composed of 35.7 g/l of sodium hexametaphosphate and 7.9 g/l of sodium carbonate.

5.5   Rate of energy dissipation

With reference to Section 3.7, the rate of energy dissipation can be expressed in terms of the stresses applied on a fluid element. The rate of work done per unit mass χ by external forces on a fluid element is the product of the gradient of stress and the respective velocity component:

$$\chi = \left[\frac{\partial \sigma_x}{\partial x} + \frac{\partial \tau_{yx}}{\partial y} + \frac{\partial \tau_{zx}}{\partial z}\right]v_x$$
$$+ \left[\frac{\partial \tau_{xy}}{\partial x} + \frac{\partial \sigma_y}{\partial y} + \frac{\partial \tau_{zy}}{\partial z}\right]v_y + \left[\frac{\partial \tau_{xz}}{\partial x} + \frac{\partial \tau_{yz}}{\partial y} + \frac{\partial \sigma_z}{\partial z}\right]v_z \tag{5.25}$$

Expanding these terms gives

$$\chi = \left[\frac{\partial}{\partial x}(\sigma_x v_x + \tau_{yx} v_y + \tau_{zx} v_z)\right.$$
$$\left. + \frac{\partial}{\partial y}(\tau_{xy} v_x + \sigma_y v_y + \tau_{zy} v_z) + \frac{\partial}{\partial z}(\tau_{xz} v_x + \tau_{yz} v_y + \sigma_z v_z)\right]$$
$$- \left(\sigma_x \frac{\partial v_x}{\partial x} + \sigma_y \frac{\partial v_y}{\partial y} + \sigma_z \frac{\partial v_z}{\partial z}\right) - \tau_{zx}\left(\frac{\partial v_z}{\partial x} + \frac{\partial v_x}{\partial z}\right)$$

Table 5.5. *Dissipation function* χ_D *for incompressible Newtonian fluids*

Cartesian

$$\chi_D = \mu_m \left\{ 2\left[\left(\frac{\partial v_x}{\partial x}\right)^2 + \left(\frac{\partial v_y}{\partial y}\right)^2 + \left(\frac{\partial v_z}{\partial z}\right)^2\right] + \left[\frac{\partial v_y}{\partial x} + \frac{\partial v_x}{\partial y}\right]^2 + \left[\frac{\partial v_z}{\partial y} + \frac{\partial v_y}{\partial z}\right]^2 + \left[\frac{\partial v_x}{\partial z} + \frac{\partial v_z}{\partial x}\right]^2 \right\}$$

Cylindrical

$$\chi_D = \mu_m \left\{ 2\left[\left(\frac{\partial v_r}{\partial r}\right)^2 + \left(\frac{1}{r}\frac{\partial v_\theta}{\partial \theta} + \frac{v_r}{r}\right)^2 + \left(\frac{\partial v_z}{\partial z}\right)^2\right] \right. $$
$$\left. + \left[r\frac{\partial}{\partial r}\left(\frac{v_\theta}{r}\right) + \frac{1}{r}\frac{\partial v_r}{\partial \theta}\right]^2 + \left[\frac{1}{r}\frac{\partial v_z}{\partial \theta} + \frac{\partial v_\theta}{\partial z}\right]^2 + \left[\frac{\partial v_r}{\partial z} + \frac{\partial v_z}{\partial r}\right]^2 \right\}$$

Spherical

$$\chi_D = \mu_m \left\{ 2\left[\left(\frac{\partial v_r}{\partial r}\right)^2 + \left(\frac{1}{r}\frac{\partial v_\theta}{\partial \theta} + \frac{v_r}{r}\right)^2 + \left(\frac{1}{r\sin\theta}\frac{\partial v_\varphi}{\partial \varphi} + \frac{v_r}{r} + \frac{v_\theta \cot\theta}{r}\right)^2\right] + \left[r\frac{\partial}{\partial r}\left(\frac{v_\theta}{r}\right) + \frac{1}{r}\frac{\partial v_r}{\partial \theta}\right]^2 \right. $$
$$\left. + \left[\frac{\sin\theta}{r}\frac{\partial}{\partial \theta}\left(\frac{v_\varphi}{\sin\theta}\right) + \frac{1}{r\sin\theta}\frac{\partial v_\theta}{\partial \varphi}\right]^2 + \left[\frac{1}{r\sin\theta}\frac{\partial v_r}{\partial \varphi} + r\frac{\partial}{\partial r}\left(\frac{v_\varphi}{r}\right)\right]^2 \right\}$$

$$-\tau_{zy}\left(\frac{\partial v_z}{\partial y} + \frac{\partial v_y}{\partial z}\right) - \tau_{xy}\left(\frac{\partial v_x}{\partial y} + \frac{\partial v_y}{\partial x}\right) \tag{5.26}$$

The terms in brackets on the first line of Equation (5.26) describe the rate of increase in mechanical energy of the fluid and are not dissipative. For Newtonian mixtures, the last four terms of the equation describe the rate of energy dissipation per unit mass χ_D, which from Equations (5.2) and (5.3) can be rewritten as

$$\chi_D = -p\left(\frac{\partial v_x}{\partial x} + \frac{\partial v_y}{\partial y} + \frac{\partial v_z}{\partial z}\right) + 2\mu_m\left[\left(\frac{\partial v_x}{\partial x}\right)^2 + \left(\frac{\partial v_y}{\partial y}\right)^2 + \left(\frac{\partial v_z}{\partial z}\right)^2\right]$$
$$-\frac{2}{3}\mu_m\left[\frac{\partial v_x}{\partial x} + \frac{\partial v_y}{\partial y} + \frac{\partial v_z}{\partial z}\right]^2$$
$$+\mu_m\left[\left(\frac{\partial v_y}{\partial x} + \frac{\partial v_x}{\partial y}\right)^2 + \left(\frac{\partial v_z}{\partial y} + \frac{\partial v_y}{\partial z}\right)^2 + \left(\frac{\partial v_x}{\partial z} + \frac{\partial v_z}{\partial x}\right)^2\right] \tag{5.27}$$

The first term of Equation (5.27) expresses the rate at which the fluid is compressed and vanishes for incompressible fluids. The last three terms in brackets involve the fluid viscosity μ_m, and their sum determines the rate at which energy is dissipated through viscous action per unit volume of fluid. The dissipation χ_D for incompressible Newtonian fluids in Cartesian, cylindrical, and spherical coordinates is given in Table 5.5. In Cartesian coordinates, the dissipation function then reduces to

$$\chi_D = 2\mu_m\left[\left(\frac{\partial v_x}{\partial x}\right)^2 + \left(\frac{\partial v_y}{\partial y}\right)^2 + \left(\frac{\partial v_z}{\partial z}\right)^2\right] + \mu_m[\mathcal{O}_x^2 + \mathcal{O}_y^2 + \mathcal{O}_z^2] \tag{5.28}$$

Energy is dissipated through linear deformation [first term in brackets in Eq. (5.28)] and through angular deformation [second term in brackets in Eq. (5.28)].

Example 5.2   Analysis of viscous energy dissipation around a sphere. Calculate the total rate of energy dissipation X_D for flow around a sphere of diameter d_s in creeping motion at a velocity u_∞ in an infinite mass of fluid. The dissipation function in spherical coordinates χ_D is integrated outside a sphere of radius $R = d_s/2$, on the fluid element $dr \times r\, d\theta \times r\sin\theta\, d\varphi$:

$$X_D = \int_0^{2\pi}\int_0^{\pi}\int_{d_s/2}^{r} \chi_D\, r^2\, dr\, \sin\theta\, d\theta\, d\varphi \tag{E5.2.1}$$

For incompressible flow around the sphere ($v_\varphi = 0$) the dissipation function in spherical coordinates from Table 5.5 reduces to

$$X_D = 2\pi\mu_m \int_0^{\pi}\int_{d_s/2}^{r}\left\{2\left(\frac{\partial v_r}{\partial r}\right)^2 + 2\left(\frac{1}{r}\frac{\partial v_\theta}{\partial \theta} + \frac{v_r}{r}\right)^2 + 2\left(\frac{v_r}{r} + \frac{v_\theta \cot\theta}{r}\right)^2\right.$$
$$\left. + \left[r\frac{\partial}{\partial r}\left(\frac{v_\theta}{r}\right) + \frac{1}{r}\frac{\partial v_r}{\partial \theta}\right]^2\right\} r^2 \sin\theta\, dr\, d\theta$$

which from the velocity profile in Equation (5.8) yields

$$X_D = 3\pi\mu_m u_\infty^2 d_s\left[1 - \frac{3d_s}{4r} + \frac{d_s^3}{8r^3} - \frac{d_s^5}{64r^5}\right]$$

This function, plotted in Figure E5.2.1, shows that 50% of the energy is dissipated within three times the particle diameter. The total energy dissipated at $r = \infty$ is

$$X_D = 3\pi\mu_m u_\infty^2 d_s = F_D u_\infty \qquad \blacklozenge$$

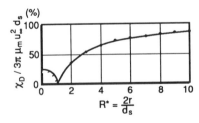

Figure E5.2.1. Total energy dissipated around a sphere

The total rate of energy dissipated per unit volume around the sphere ∀ is

$$\frac{X_D}{\forall} = \frac{18\pi\mu_m u_\infty^2 d_s}{\pi d_s^3} = 18\mu_m\left(\frac{u_\infty}{d_s}\right)^2 \qquad (E5.2.2)$$

This equation indicates that, for creeping motion at constant velocity u_∞, the total rate of energy dissipation per unit volume of the sphere is inversely proportional to d_s^2. On the other hand, when u_∞ corresponds to the fall velocity ω [Eq. (5.21)], the rate of energy dissipation becomes proportional to d_s^2 for particles finer than silts [from (Eq. 5.21)] and inversely proportional to d_s for particles coarser than gravels [from Eq. (5.22)].

5.6   Laboratory measurements of particle size

Two principal functions of a sediment laboratory are to determine (1) particle size distribution of suspended-sediment and streambed material and (2) concentration of suspended sediment. Other functions include the determination of roundness and shape of individual grains and their mineral composition, the amount of organic matter, the specific gravity of sediment particles, and the specific weight of deposits.

There are essentially two ways to determine particle size distributions in the laboratory: direct measurement and sedimentation methods. The direct methods, also discussed in Section 2.3, include immersion and displacement volume measurements, some direct measurements of circumference or diameter, and semidirect measurements of particle diameter using sieves. Sedimentation methods relate fall velocity measurements to particle size. Standard devices include the visual accumulation tube (VAT), the bottom-withdrawal tube (BWT), the pipette, and the hydrometer. The VAT, used only for sands, operates as a stratified system where particles start from a common source at the top and deposit at the bottom of the tube according to settling velocities. The pipette and BWT, used only for silts and clays, operate as dispersed systems where particles begin to settle from an initially uniform dispersion. For the BWT method, the distribution is obtained from the quantity of sediment remaining in suspension after various settling times when the coarser sizes and heavier concentrations are withdrawn from the bottom of the tube. Table 5.6 indicates recommended size ranges, analysis concentration, and weight of sediment for the sieve, VAT, BWT, and pipette methods.

Table 5.6. *Recommended characteristics of particle size analysis*

Method of analysis	Recommended size range (mm)	Desirable concentration (mg/l)	Optimum sediment quantity (g)
Sieves	0.062–32	—	1 g to 10 kg
VAT	0.062–2.0	—	0.05–15.0
Pipette	0.002–0.062	2,000–5,000	1.0–5.0
BWT	0.002–0.062	1,000–3,500	0.5–1.8

Extraneous organic materials should be removed from samples by adding about 5 ml of 6% solution of hydrogen peroxide for each gram of dry sample in 40 ml of water. The solution must be stirred thoroughly and covered for about 10 min. Large fragments of organic material may then be skimmed off when they are free of sediment particles. If oxidation is slow, or after it has slowed, the mixture is heated to 93°C, stirred occasionally, and more hydrogen peroxide solution added as needed. After the reaction has stopped, the sediment must be carefully washed two or three times with distilled water.

To ensure complete dispersion of silts and clays when using the pipette, BWT, and hydrometer methods, 1 ml of dispersing agent should be used for each 100-ml sample. Adding 35.7 g/l of sodium hexametaphosphate and 7.99 g/l of sodium carbonate is recommended to prevent flocculation (Section 5.4.3).

The following indirect methods of particle size measurement involve liquid suspensions and fall velocity: the VAT, BWT, pipette, and hydrometer methods.

5.6.1   *Visual accumulation tube method*

The VAT method is a fast, economical, and reasonably accurate means of determining the sand size distribution based on fall velocity measurements. Silts finer than 0.062 mm are removed either by wet sieving or by sedimentation methods and analyzed separately using either the pipette or the BWT method. In some instances, sieving must be employed to remove particles coarser than 2.0 mm.

The equipment for the VAT method of analysis consists primarily of a special settling tube and a recording mechanism, as shown in Figure 5.3.

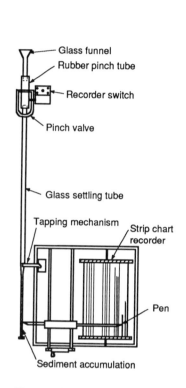

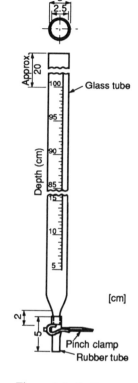

Figure 5.3. Visual accumulation tube

Figure 5.4. Bottom-withdrawal tube

The VAT analysis results in a continuous pen trace of sediment accumulation as a function of time. The chart is calibrated at a given temperature to determine the relative amount of sediment in terms of fall diameter and percent finer than a given size.

The sediment size distribution is corrected for the finer or coarser fractions removed before the VAT analysis. For instance, if 30% of the original sample finer than 0.062 mm was removed from the VAT analysis, the percent finer scale of the VAT analysis is corrected to start at 30%.

5.6.2 Bottom-withdrawal tube method

The bottom-withdrawal tube (Fig. 5.4) is used when the sample contains a very small quantity of fine sands and silts. Fractions coarser than 0.0625 mm should be removed and analyzed separately with the VAT.

Table 5.7. *BWT withdrawal time in minutes*

Temp. (°C)	Particle diameter (mm)							
	0.25	0.125	0.0625	0.0312	0.0156	0.0078	0.0039	0.00195
18	0.522	1.48	5.02	20.1	80.5	322	1,288	5,154
19	0.515	1.45	4.88	19.6	78.5	314	1,256	5,026
20	0.508	1.41	4.77	19.2	76.6	306	1,225	4,904
21	0.503	1.39	4.67	18.7	74.9	299	1,198	4,794
22	0.497	1.37	4.55	18.3	73.0	292	1,168	4,675
23	0.488	1.34	4.45	17.8	71.3	285	1,141	4,566
24	0.485	1.32	4.33	17.4	69.6	279	1,114	4,461
25	0.478	1.30	4.25	17.0	68.1	273	1,090	4,361
26	0.472	1.28	4.15	16.7	66.6	266	1,065	4,263
27	0.467	1.26	4.05	16.3	65.1	260	1,042	4,169
28	0.462	1.24	3.97	15.9	63.7	255	1,019	4,079
29	0.455	1.22	3.88	15.6	62.3	249	997	3,991
30	0.450	1.20	3.80	15.3	61.0	244	976	3,907
31	0.445	1.18	3.71	14.9	59.7	239	956	3,825
32	0.442	1.17	3.65	14.6	58.5	234	936	3,747
33	0.438	1.15	3.58	14.2	57.3	229	917	3,671
34	0.435	1.13	3.51	13.9	56.1	224	898	3,494

Note: Time in minutes required for spheres having a specific gravity of 2.65 to fall 1 m in water at varying temperatures.

A 100-ml suspension is poured into a 1-m-high settling tube, and 10-ml samples are withdrawn at the lower end of the tube following the schedule given in Table 5.7. The samples are then poured into evaporating dishes and the sample containers washed with distilled water. The previously weighted evaporating dishes are placed in the oven to dry at a temperature just below the boiling point, to avoid splattering by boiling. When the evaporating dishes or flasks are visibly dry, the temperature is raised to 110° for 1 h, after which the containers are transferred from the oven to a desiccator and allowed to cool before weight measurements are made. The sediment size distribution is then calculated using the Oden curve method.

The Oden curve in Figure 5.5 shows the percentage by weight of sediment in suspension as a function of time. If tangents are drawn to the Oden curve at any two consecutive times of withdrawal t_i and t_{i+1} (Table 5.7), the tangents then intersect the ordinate at %W_i and %W_{i+1}, respectively. The difference between %W_{i+1} and %W_i represents the percentage by weight of material in the size range corresponding to the settling times

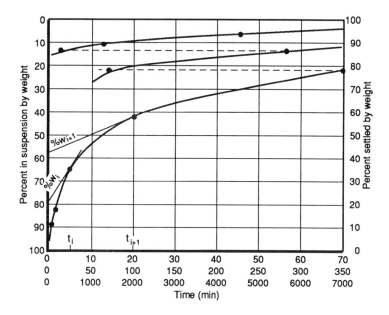

Figure 5.5. Example of an Oden curve

t_i and t_{i+1}. Obviously, the values $\%W_i$ and W_{i+1} are the percentages by weight of the sediment in the sample that are finer than sizes corresponding to t_i and t_{i+1}, respectively. For instance, Figure 5.5 shows that the sediment sample contains $78\% - 58\% = 20\%$ of coarse silt (0.0312 mm $<$ $d_s < 0.0625$ mm); see Problem 5.4 for details.

Care is needed in drawing the tangents to the Oden curve, as the curvature greatly affects the position of the intercept on the percentage scale. For samples containing clays, the slope of the curve does not approach zero at the time of the last scheduled withdrawal because clay particles are still settling. Obviously, the Oden curve of dispersed suspensions should never have a reverse slope.

5.6.3 Pipette method

The pipette method is a reliable indirect method for determining the particle size distribution of silts and coarse clays ($d_s < 0.062$ mm). The concentration of a quiescent suspension is measured at a predetermined depth (Fig. 5.6) as a function of settling time. Particles having a settling velocity greater than a given size settle below the point of withdrawal after a certain time, as determined in Table 5.8. The time and depth of withdrawal

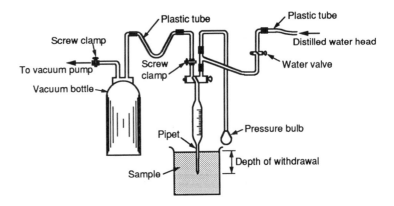

Figure 5.6. A pipette

are predetermined on the basis of Stokes' law [Eq. (5.21)] at a given water temperature. Samples are dried and weighed, and the Oden curve method is used to calculate the particle size distribution.

5.6.4   Hydrometer method

The hydrometer measures the change in immersed volume of a floating object in a dilute suspension. The buoyancy force from Equation (E3.3.1) being equal to the weight of the object, the time change in submerged volume corresponds to the time change in specific weight of the mixture due to settling of particles in suspension.

Exercises

5.1.   Demonstrate that $\tau_{xy} = \tau_{yx}$ from the sum of moments about the center of an infinitesimal element.

*5.2.   Derive the x component of the Navier–Stokes equations in Table 5.1 from the equation of motion [Eq. (3.12a)] and the stress tensor components for incompressible Newtonian fluids [Eqs. (5.2) and (5.3)].

*5.3.   Cross-differentiate Equations (5.4a) and (5.4b) along y and x, respectively, to derive Equation (5.5c).

5.4.   Use the definition of stream functions in Cartesian coordinates [Eq. (4.1)] to demonstrate that Equation (5.6) results from Equations (5.5).

Table 5.8. *Pipette withdrawal time*

Temp. (°C)	0.062; 15	0.031; 15		0.016; 10		0.008; 10		0.004; 5		0.002; 5	
			Diameter of particles (mm); depth of withdrawal (cm)								
20	44 s	2 m	52 s	7 m	40 s	30 m	40 s	61 m	19 s	4 h	5 m
21	42	2	48	7	29	29	58	59	50	4	0
22	41	2	45	7	18	29	13	58	22	3	54
23	40	2	41	7	8	28	34	57	5	3	48
24	39	2	38	6	58	27	52	55	41	3	43
25	38	2	34	6	48	27	14	54	25	3	38
26	37	2	30	6	39	26	38	53	12	3	33
27	36	2	27	6	31	26	2	52	2	3	28
28	36	2	23	6	22	25	28	50	52	3	24
29	35	2	19	6	13	24	53	49	42	3	19
30	34	2	16	6	6	24	22	48	42	3	15

Note: The values of withdrawal time in this table (given in minutes and seconds) are based on particles of assumed spherical shape with an average specific gravity of 2.65, the constant of acceleration due to gravity = 980 cm/s^2, and viscosity varying from 0.010087 cm^2/s at 20°C to 0.008004 cm^2/s at 30°C.

*5.5. Determine the shear stress component $\tau_{r\theta}$ in Equation (5.9c) from the tensor $\tau_{r\theta}$ in spherical coordinates (Table 5.3) and the velocity relationships [Eqs. (5.8a) and (5.8b)].

**5.6. (a) Integrate the shear stress distribution [Eq. (5.9c)] to determine the surface drag in Equation (5.11) from Equation (5.10); and

(b) integrate the dynamic pressure distribution [Eq. (5.9b)] to obtain the form drag in Equation (5.13) from Equation (5.12).

*5.7. Derive Rubey's fall velocity equation [Eqs. (5.23a) and (5.23b)] combining Equations (5.18) and (5.22b).

5.8. Substitute the appropriate stress tensor components for the flow of Newtonian fluids in Cartesian coordinates [Eqs. (5.2) and (5.3)] into the last four terms in parentheses of Equation (5.26) to obtain the energy dissipation function in Equation (5.27).

5.9. Describe each member of the dissipation function [Eq. (5.28)] in terms of the fundamental modes of deformation shown in Figure 3.2 (translation, linear deformation, angular deformation, and rotation of a fluid element).

**Problem 5.1

Plot the velocity, dynamic pressure, and shear stress distributions around
the surface of a sphere for creeping motion given by Stokes' law [Eqs.
(5.8) and (5.9)] and compare with irrotational flow without circulation
(Problem 4.3).

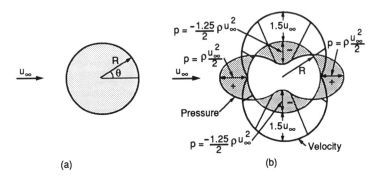

(a)   (b)

(a) Creeping motion in a viscous fluid

$v = 0$ everywhere at the sphere surface
$\tau \neq 0$ everywhere at the sphere surface except at $\theta = 0°$
and $180°$
$p_d \neq 0$ everywhere at the sphere surface except at $\theta = 90°$

(b) Irrotational flow

$\tau = 0$ everywhere at the sphere surface
$v \neq 0$ everywhere at the sphere surface except at $\theta = 0°$
and $180°$
$p_d \neq 0$ everywhere at the sphere surface except at $\theta = 41.8°$
and $138.2°$

*Problem 5.2

Plot Rubey's relationship for the drag coefficient C_D in Figure 5.2. How
does it compare with the experimental measurements? At a given Re_p,
which of Equations (5.18) and (5.19) induces larger settling velocities?

*Problem 5.3

Evaluate the dissipation function χ_D from Table 5.5 for a vertical axis
Rankine vortex described in cylindrical coordinates by

(a) forced vortex

$$v_\theta = \frac{\Gamma_v r}{2\pi r_0^2}, \qquad v_z = v_r = 0 \qquad \text{(rotational flow for } r < r_0\text{)}$$

and

(b) free vortex

$$v_\theta = \frac{\Gamma_v}{2\pi r}, \qquad v_z = v_r = 0 \qquad \text{(irrotational flow for } r > r_0\text{)}$$

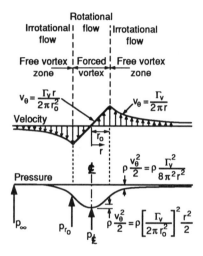

Answer

$$\chi_D = \mu_m \left[r \frac{\partial}{\partial r} \left(\frac{v_\theta}{r} \right) \right]^2$$

(a) $\chi_D = 0$, although the flow is rotational for $r < r_0$; and
(b) $\chi_D = \mu_m \Gamma_v^2 / \pi^2 r^4 \neq 0$, although the flow is irrotational for $r > r_0$.

**Problem 5.4

The sediment size distribution of a 1,200-mg sample is to be determined using the BWT. If the water temperature is 24°C and the solid weight for each 10-ml withdrawal is given, complete the following table and determine the particle size distribution:

Particle diameter (m)	Withdrawal time (min)	Sample volume (ml)	Dry weight of sediment (mg)	Cumulative dry weight (mg)	Percent settled	Percent finer[a] (%W)
0.25	0.485	10	144	144	12	
0.125		10	72	216	18	
0.0625		10	204	420	35	78
0.0312		10	264	684	57	58
0.0156		10	252			
0.0078		10	84		84	
0.0039		10	48	1,068		
0.00195	4,461	10	45			

[a] See Figure 5.5.

**Computer Problem 5.1

Write a simple computer program to determine the particle size d_s, the fall velocity ω, the flocculated fall velocity ω_f, the particle Reynolds number Re_p, the dimensionless particle diameter d_*, and the time of settling per meter of quiescent water at 5°C, and complete the following table:

Class name	d_s (mm)	ω (cm/s)	ω_f (cm/s)	Re_p	d_*	Settling time
Medium clay						
Medium silt						
Medium sand						
Medium gravel						
Small cobble						
Medium boulder						

6

Turbulent velocity profiles

Most sediment-laden flows are characterized by irregular velocity fluctuations indicating turbulence. The turbulent fluctuation superimposed on the principal motion is complex and remains difficult to treat mathematically. This chapter outlines the fundamentals of turbulence with emphasis on turbulent velocity profiles (Section 6.1), turbulent flow along rough and smooth boundaries (Sections 6.2 and 6.3), departure from logarithmic velocity profiles (Section 6.4), and open-channel flow measurements (Section 6.5).

In describing flow in mathematical terms, it is convenient to separate the mean motion (notation with overbar) from the fluctuation (notation with superscript +) as sketched in Figure 6.1. Denoting a fluctuating parameter $\hat{v}_x$ of time-averaged value $\bar{v}_x$ and fluctuation v_x^+, the pressure and the velocity components can be rewritten as

$$\hat{p} = \bar{p} + p^+ \tag{6.1a}$$

$$\hat{v}_x = \bar{v}_x + v_x^+ \tag{6.1b}$$

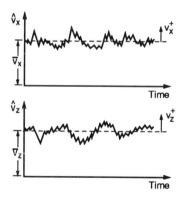

Figure 6.1. Time velocity measurements

$$\hat{v}_y = \bar{v}_y + v_y^+ \tag{6.1c}$$

$$\hat{v}_z = \bar{v}_z + v_z^+ \tag{6.1d}$$

The time-averaged values at a fixed point in space are given by

$$v_x = \bar{v}_x = \frac{1}{t_1} \int_{t_0}^{t_0+t_1} \hat{v}_x \, dt \tag{6.2}$$

Taking the mean values over a sufficiently long time interval t_1, the time-averaged values of the fluctuations equal zero; thus, $\overline{v_x^+} = \overline{v_y^+} = \overline{v_z^+} = \overline{p^+} = 0$.

Likewise, the time-averaged values of the derivatives of velocity fluctuations, such as $\overline{\partial v_x^+/\partial x}$, $\overline{\partial^2 v_x^+/\partial x^2}$, $\overline{\partial \bar{v}_x v_x^+/\partial x^2}$, also vanish, owing to Equation (6.2). The quadratic terms arising from the products of cross-velocity fluctuations such as $\overline{v_x^+ v_x^+}$, $\overline{v_x^+ v_y^+}$, $\overline{\partial v_x^+ v_y^+/\partial x}$, however, do not vanish. The overbar of simple time-averaged parameters is omitted for notational convenience.

It is seen that both the time-averaged velocity components and the fluctuating components satisfy the equation of continuity. Thus, for incompressible fluids,

$$\frac{\partial v_x}{\partial x} + \frac{\partial v_y}{\partial y} + \frac{\partial v_z}{\partial z} = 0 \tag{6.3a}$$

$$\frac{\partial v_x^+}{\partial x} + \frac{\partial v_y^+}{\partial y} + \frac{\partial v_z^+}{\partial z} = 0 \tag{6.3b}$$

The velocity and pressure terms from Equation (6.1) are substituted into the Navier–Stokes equations (Table 5.1) to give the following acceleration terms:

$$\frac{\partial v_x}{\partial t} + v_x \frac{\partial v_x}{\partial x} + v_y \frac{\partial v_x}{\partial y} + v_z \frac{\partial v_x}{\partial z} = g_x - \frac{1}{\rho_m} \frac{\partial p}{\partial x} + \nu_m \nabla^2 v_x - \left[\frac{\overline{\partial v_x^+ v_x^+}}{\partial x} + \frac{\overline{\partial v_y^+ v_x^+}}{\partial y} + \frac{\overline{\partial v_z^+ v_x^+}}{\partial z} \right] \tag{6.4a}$$

$$\frac{\partial v_y}{\partial t} + v_x \frac{\partial v_y}{\partial x} + v_y \frac{\partial v_y}{\partial y} + v_z \frac{\partial v_y}{\partial z} = g_y - \frac{1}{\rho_m} \frac{\partial p}{\partial y} + \nu_m \nabla^2 v_y - \left[\frac{\overline{\partial v_x^+ v_y^+}}{\partial x} + \frac{\overline{\partial v_y^+ v_y^+}}{\partial y} + \frac{\overline{\partial v_z^+ v_y^+}}{\partial z} \right] \tag{6.4b}$$

$$\frac{\partial v_z}{\partial t} + v_x \frac{\partial v_z}{\partial x} + v_y \frac{\partial v_z}{\partial y} + v_z \frac{\partial v_z}{\partial z} = g_z - \frac{1}{\rho_m} \frac{\partial p}{\partial z} + \nu_m \nabla^2 v_z - \left[\frac{\overline{\partial v_x^+ v_z^+}}{\partial x} + \frac{\overline{\partial v_y^+ v_z^+}}{\partial y} + \frac{\overline{\partial v_z^+ v_z^+}}{\partial z} \right] \tag{6.4c}$$

| local | convective | grav-ita-tional | pressure gradient | viscous | turbulent fluctuations |

In addition to the terms found in the Navier–Stokes equations, three cross-products of velocity fluctuations are obtained from the convective acceleration terms on the left-hand side of Equation (6.4). These turbulent acceleration terms provide additional stresses called Reynolds stresses, which are usually added to the right-hand side of Equation (6.4). Generally speaking, these turbulent acceleration terms far outweigh the viscous components in turbulent flow.

6.1   Logarithmic velocity profiles

Since the fluid does not slip at solid boundaries, all turbulent components must vanish at the walls and remain very small in their immediate neighborhood. It follows that, near the boundary, all turbulent acceleration terms in Equation (6.4) become smaller than the viscous acceleration terms of the Navier–Stokes equations. In turbulent flows, laminar motion must therefore persist in a very thin layer next to the boundary. This is known as the laminar sublayer.

Consider a thin flat plate set parallel to the main flow direction x. We are interested in describing the time-averaged velocity profile v_x as a function of the distance z away from the plate. Drawing an analogy with the mean free path in the kinetic theory of gases, Prandtl imagined the mixing-length concept, which implies that the transverse velocity fluctuation v_z^+ is of the same order of magnitude as v_x^+. He hypothesized that the average of the absolute value of velocity fluctuations is proportional to the velocity gradient in the form

$$\overline{|v_x^+|} \sim \overline{|v_z^+|} \sim l_{\mathrm{m}} \frac{dv_x}{dz} \tag{6.5}$$

in which the proportionality constant l_{m} denotes the Prandtl mixing length. The average products of velocity fluctuations were then formulated in terms of the mixing length with the aid of equation (6.5):

$$\overline{v_x^+ v_z^+} \sim -\overline{|v_x^+|}\,\overline{|v_z^+|} \tag{6.6a}$$

$$\overline{v_x^+ v_z^+} \sim -l_{\mathrm{m}}^2 \left(\frac{dv_x}{dz}\right)^2 \tag{6.6b}$$

The corresponding shear stress of a mixture can then be written as

$$\tau_{zx} = \underbrace{\mu_{\mathrm{m}} \frac{dv_x}{dz}}_{\text{viscous}} + \underbrace{\rho_{\mathrm{m}} l_{\mathrm{m}}^2 \left(\frac{dv_x}{dz}\right)^2}_{\text{turbulent}} \tag{6.7}$$

In turbulent flows, the turbulent shear stress far outweighs the viscous shear stress. The converse is true in the laminar sublayer. The turbulent shear stress can be written alternatively as a function of the Boussinesq eddy viscosity ϵ_m:

$$\tau_{zx} \cong \rho_m \epsilon_m \frac{dv_x}{dz} = \rho_m l_m^2 \left[\frac{dv_x}{dz} \right]^2 \tag{6.8}$$

It is further assumed that the mixing length l_m is proportional to the distance z from the boundary,

$$l_m = \kappa z \tag{6.9}$$

in which κ is the von Kármán constant ($\kappa \simeq 0.4$).

The shear stress τ_{zx} in the region close to the wall remains virtually equal to the boundary shear stress $\tau_0 = \rho u_*^2$; thus, for the turbulent region near the boundary one obtains the following equation for the shear velocity u_* from Equations (6.8) and (6.9) when $\tau_{zx} = \tau_0 = \rho_m u_*^2$:

$$\sqrt{\frac{\tau_0}{\rho_m}} = u_* = \kappa z \left(\frac{dv_x}{dz} \right) \tag{6.10}$$

Since u_* is constant, the variables v_x and z can be separated and integrated to yield the logarithmic average velocity distribution for steady turbulent flow near a flat boundary,

$$\frac{v_x}{u_*} = \frac{1}{\kappa} \ln z + c_0 \tag{6.11}$$

in which c_0 is an integration constant evaluated at a distance z_0 from the flat boundary, where the logarithmic velocity v_{x0} hypothetically equals zero. Hence,

$$\frac{v_x}{u_*} = \frac{1}{\kappa} \ln \frac{z}{z_0} \tag{6.12}$$

Two types of boundary conditions are recognized depending on the relative magnitude of the grain size d_s and the laminar sublayer thickness δ, examined in Section 6.3. Conceptually, the boundary is said to be hydraulically smooth when $\delta \gg d_s$ and, conversely, hydraulically rough when $d_s \gg \delta$ (Section 6.2). A transition zone is also recognized, as shown in Figure 6.2.

6.2   Rough plane boundary

Consider steady turbulent flow over a plane surface of coarse solid particles of grain roughness height k_s'. The flow is described as turbulent over

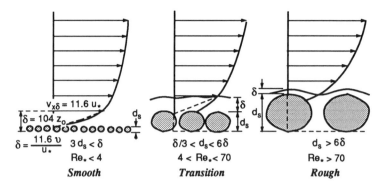

Figure 6.2. Hydraulically smooth and rough boundaries

a rough boundary when $k_s' \gg \delta$. Gravel-bed and cobble-bed streams are considered hydraulically rough. Early experiments in pipes indicated that in such a case the distance $z_0 = k_s'/30$ and the corresponding velocity profile is

$$\frac{v_x}{u_*} = \frac{2.3}{\kappa} \log\left(\frac{z}{k_s'}\right) + 8.5 = 5.75 \log\left(\frac{30z}{k_s'}\right) \qquad (6.13) \blacklozenge$$

which yields the depth-averaged velocity V_x for wide rectangular channels after integration of Equation (6.13) over the entire flow depth h:

$$\frac{V_x}{u_*} = \frac{2.3}{\kappa} \log\left(\frac{h}{k_s'}\right) + 6.25 = 5.75 \log\left(\frac{12.2h}{k_s'}\right) \qquad (6.14) \blacklozenge$$

Note that the integration of Equation (6.13) strictly yields an integration constant of 6.0 in Equation (6.14), while the given value 6.25 is commonly referenced.

The total resistance to flow can be described in terms of the Chézy coefficient C, the Darcy–Weisbach friction factor f, or the Manning coefficient n. The following identity between these three factors has been established:

$$C \equiv \sqrt{\frac{8g}{f}} \equiv \frac{R_h^{1/6}}{n} \text{ (in SI units)}$$

$$\equiv \frac{1.49 R_h^{1/6}}{n} \text{ (in English units)} \qquad (6.15a) \blacklozenge$$

where R_h is the hydraulic radius and g the gravitational acceleration. The fundamental dimensions are as follows: C is in $L^{1/2}/T$, f is dimensionless, and n is in $T/L^{1/3}$.

In plane bed channels, the total resistance is composed solely of grain resistance (indicated by a prime), and the corresponding grain resistance parameters are (1) the grain Chézy coefficient C'; (2) the Darcy–Weisbach grain friction factor f'; and (3) the grain Manning coefficient n'. The identity between these grain resistance parameters is

$$C' \equiv \sqrt{\frac{8g}{f'}} \equiv \frac{R_h^{1/6}}{n'} \text{ (in SI units)}$$

$$\equiv \frac{1.49R_h^{1/6}}{n'} \text{ (in English units)} \tag{6.15b}$$

In plane bed channels, the total resistance equals the grain resistance, and $\tau_0 = \tau_0'$, $u_* = u_*'$, $C = C'$, $f = f'$, and $n = n'$.

For plane bed channels, depth-averaged velocity relationships [e.g., Eq. (6.14)] directly express grain resistance to flow in terms of the Darcy–Weisbach grain friction factor f' through the following identities:

$$\tau_0' \equiv \frac{f'}{8}\rho_m V_x^2 \tag{6.16a}$$

or

$$u_*'^2 \equiv \frac{f'}{8}V_x^2 \tag{6.16b}$$

The corresponding depth-averaged velocity relationships and their range of applicability are compiled in Table 6.1. For sand-bed channels, Kamphuis (1974) recommended $k_s' = 2d_{90}$. For gravel-bed rivers, Bray (1982) found $k_s' = 3.1d_{90}$; $k_s' = 3.5d_{84}$; $k_s' = 5.2d_{65}$; $k_s' = 6.8d_{50}$. The relationship $k_s' \cong 3d_{90}$ appears to be a reasonably good approximation.

The resulting Darcy–Weisbach grain friction factor f' and the grain Chézy coefficient C' are commonly approximated by

$$C' \equiv \sqrt{\frac{8g}{f'}} = \frac{2.3}{\kappa}\sqrt{g}\log\left(\frac{12.2R_h}{3d_{90}}\right) \cong 5.75\sqrt{g}\log\left(\frac{4h}{d_{90}}\right) \tag{6.17}$$

It follows that grain resistance (f', C', or n') for turbulent flow over a rough boundary can be obtained from (1) the depth-averaged velocity V_x and grain shear stress τ_0' from Equation (6.16a); (2) the grain shear velocity u_*' and the depth-averaged velocity V_x from Equation (6.16b); or (3) the flow depth h and the grain size d_{90} from Equation (6.17).

The grain resistance equation in logarithmic form can be transformed into an equivalent power form in which the exponent b varies with relative submergence h/d_s,

Table 6.1. *Grain resistance and velocity formulations for turbulent flow over hydraulically rough plane boundaries* ($C = C'$ *and* $f = f'$)

Formulation	Range	Resistance parameter	Velocity[a]
Chézy	$h/d_s \to \infty$	$C = \sqrt{\dfrac{8g}{f}}$ constant	$V = CR_h^{1/2}S_f^{1/2}$
Manning	$h/d_s > 100$	$C \cong a\left(\dfrac{R_h}{d_s}\right)^{1/6} \cong \dfrac{R_h^{1/6}}{n}$ (SI)	$V = \dfrac{1}{n}R_h^{2/3}S_f^{1/2}$ (SI)
			$n \cong 0.062d_{50}^{1/6}$ (d_{50} in m)
			$n \cong 0.046d_{75}^{1/6}$ (d_{75} in m)
			$n \cong 0.038d_{90}^{1/6}$ (d_{90} in m)
Logarithmic		$\dfrac{C}{\sqrt{g}} = \sqrt{\dfrac{8}{f}} = 5.75\log\left(\dfrac{12.2R_h}{k_s'}\right)$	$V = \left(5.75\log\dfrac{12.2R_h}{k_s'}\right)\sqrt{gR_hS_f}$
			$k_s' \cong 3d_{90}$
			$k_s' \cong 3.5d_{84}$
			$k_s' \cong 5.2d_{65}$
			$k_s' \cong 6.8d_{50}$

[a] The hydraulic radius $R_h = A/P$ is used, where A is the cross-sectional area and P is the wetted perimeter; the friction slope S_f is the slope of the energy grade line.

$$\sqrt{\frac{8}{f'}} = a\left(\frac{h}{d_s}\right)^b \equiv \hat{a}\ln\left[\frac{\hat{b}h}{d_s}\right] \tag{6.18}$$

under the transformation that imposes the value and the first derivative to be identical:

$$a = \frac{\hat{a}}{b}\left(\frac{d_s}{h}\right)^b \tag{6.19a}$$

and

$$b = \frac{1}{\ln(\hat{b}h/d_s)} \tag{6.19b}$$

The values of the exponent b are plotted in Figure 6.3 as a function of relative submergence h/d_s. It is shown that b gradually decreases to zero as $h/d_s \to \infty$, which implies that the Darcy–Weisbach grain friction factor f' and the grain Chézy coefficient C' are constant for very large values of h/d_s. At values of $h/d_s > 100$, the exponent b is roughly comparable to 1/6, which corresponds to the Manning–Strickler approximation ($n \sim d_s^{1/6}$). At lower values of the relative submergence $h/d_s < 100$, the

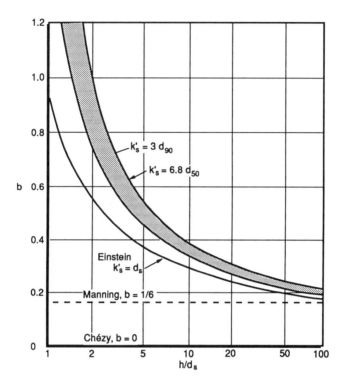

Figure 6.3. Exponent b of the grain resistance equation, $\hat{b} = 12.2$

exponent b of the power form varies with h/d_s and the logarithmic formulation is preferred.

6.3   Smooth plane boundary

Generally speaking, a plane bed surface is hydraulically smooth for grain sizes finer than medium sand ($d_s < 0.25$ mm or $d_* < 5$), as the grain roughness height k_s' becomes very small compared with the laminar sublayer thickness. For turbulent flows over a smooth boundary, the distance z_0 is proportional to the ratio ν_m/u_*, and experiments show that $z_0 \cong \nu_m/9u_*$ in smooth pipes. Substituting into the velocity profile relationship [Eq. (6.12)], one obtains

$$\frac{v_x}{u_*} = 5.75 \log\left(\frac{u_* z}{\nu_m}\right) + 5.5 \qquad (6.20) \blacklozenge$$

This relationship (plotted in Fig. 6.4) is valid for steady turbulent flow near a smooth plane boundary.

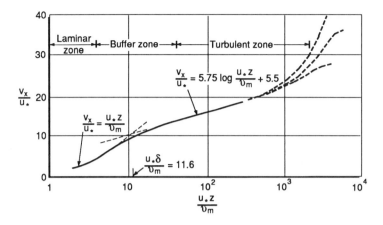

Figure 6.4. Velocity profiles for the law of the wall

The depth-averaged velocity V_x is then obtained from the integration of Equation (6.20) over the entire flow depth h. By definition of the Darcy-Weisbach grain friction factor f', from $\tau_0' = (f'/8)\rho_m V_x^2$ the mean flow velocity V_x is given by

$$\sqrt{\frac{8}{f'}} = \frac{V_x}{u_*} = 5.75 \log\left(\frac{u_* h}{\nu_m}\right) + 3.25 \qquad (6.21) \blacklozenge$$

A closer look at Equation (6.7) given Equation (6.9) indicates that, in all turbulent flows, a thin layer must exist very close to the boundary where the viscous shear stress overcomes the turbulent shear stress, because $l_m \to 0$ as $z \to 0$. This indicates that the flow becomes laminar in a layer of thickness δ adjacent to the wall called the *laminar sublayer*. Considering a thin laminar sublayer of thickness δ where the outermost velocity is $v_{x\delta}$, the boundary shear stress $\tau_0 = \rho_m u_*^2$ is, from Equation (5.1), given as:

$$\tau_0 = \mu_m \frac{v_{x\delta}}{\delta} = \rho_m u_*^2 \qquad (6.22a)$$

The corresponding dimensionless velocity from Equation (6.22a) is

$$\frac{v_{x\delta}}{u_*} = \frac{u_* \delta}{\nu_m} \qquad (6.22b)$$

Because the velocity profile is continuous, the velocity $v_x = v_{x\delta}$ at the innermost point of the turbulent velocity profile over a smooth boundary [Eq. (6.20)] must equal the velocity $v_{x\delta}$ at the outermost point of the laminar sublayer [Eq. (6.22b)]. The thickness δ of the laminar sublayer can therefore be determined by simultaneously solving Equations (6.22b) and (6.20) for δ; hence,

$$\delta = \frac{11.6\nu_m}{u_*} \tag{6.23} \blacklozenge$$

Consider a grain shear Reynolds number Re_* defined as the product of shear velocity and grain size over the kinematic viscosity of the mixture ν_m, $\text{Re}_* = u_* d_s / \nu_m$. Turbulent flows are called hydraulically smooth as long as the height of the boundary roughness characterized by the sediment size d_s remains much smaller than the laminar sublayer thickness [$3d_s < \delta$, which from Eq. (6.23) corresponds to $\text{Re}_* = u_* d_s / \nu_m < 4$]. Likewise, turbulent flows are called hydraulically rough when the grain size d_s far exceeds the laminar sublayer thickness ($d_s > 6\delta$, or $\text{Re}_* = u_* d_s / \nu_m > 70$). A transition zone exists where $\delta/3 < d_s < 6\delta$, or $4 < \text{Re}_* < 70$. Example 6.1 details typical calculation procedures for turbulent flows.

Example 6.1   Application to a turbulent velocity profile. Consider the given measured velocity profile for steady uniform turbulent flow in a wide rectangular channel, $R_h = h$ (Fig. E6.1.1). Consider two points 1 and 2 near the bed,

$$v_1 = \frac{u_*}{\kappa} \ln \frac{z_1}{k_s'}$$

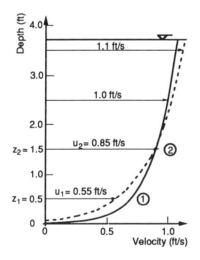

Figure E6.1.1. Measured velocity profile

$$v_2 = \frac{u_*}{\kappa} \ln \frac{z_2}{k_s'}$$

and estimate the following parameters:

(a) Shear velocity:

$$u_* = \frac{\kappa(v_2 - v_1)}{\ln(z_2/z_1)} = \frac{0.4(0.85 - 0.55)}{\ln(1.5/0.5)} = 0.11 \frac{\text{ft}}{\text{s}}$$

(b) Boundary shear stress:

$$\tau_0 = \rho_m u_*^2 = \frac{1.92 \text{ slugs}}{\text{ft}^3}(0.11)^2 \frac{\text{ft}^2}{\text{s}^2} = 0.023 \frac{\text{lb}}{\text{ft}^3}$$

(c) Laminar sublayer thickness:

$$\delta = \frac{11.6\nu_m}{u_*} = \frac{11.6 \times 1 \times 10^{-5} \text{ ft}^2\text{s}}{\text{s} \times 0.11 \text{ ft}} = 0.001 \text{ ft} = 0.32 \text{ mm}$$

The flow is hydraulically smooth if the bed material d_s is finer than about 0.1 mm or hydraulically rough if $d_s > 1.8$ mm; the transition zone roughly corresponds to sand fractions $0.1 < d_s < 1.8$ mm.

(d) Mean flow velocity:

$$V \cong 0.85 \text{ ft/s}$$

(e) Froude number:

$$\text{Fr} = \frac{V}{\sqrt{gR_h}} \cong \frac{V}{\sqrt{gh}} = \frac{0.85 \text{ ft/s}}{\sqrt{32.2 \times 3.7 \text{ ft}^2/\text{s}^2}} = 0.078$$

(f) Friction slope:

$$S_f = \frac{\tau_0}{\gamma_m R_h} \cong \frac{\tau_0}{\gamma_m h} = \frac{0.023 \text{ lb} \times \text{ft}^3}{\text{ft}^2 \times 62.4 \text{ lb} \times 3.7 \text{ ft}} = 9.96 \times 10^{-5}$$

(g) Darcy–Weisbach factor:

$$f = \frac{8S_f}{\text{Fr}^2} = \frac{8 \times 9.96 \times 10^{-5}}{0.078^2} = 0.13$$

(h) Manning coefficient:

$$n = \frac{1.49}{V} R_h^{2/3} S_f^{1/2} = \frac{1.49}{0.85}(3.7)^{2/3}(9.96 \times 10^{-5})^{1/2} = 0.042 \frac{\text{s}}{\text{ft}^{1/3}}$$

(i) Chézy coefficient:

$$C = \sqrt{\frac{8g}{f}} = \sqrt{\frac{8 \times 32.2}{0.13}} = 44.5 \, \frac{ft^{1/2}}{s}$$

(j) Momentum correction factor [Eq. (E3.5.1)]:

$$\beta_m = \frac{1}{AV_x^2} \int_A v_x^2 \, dA \cong \frac{1}{hV_x^2} \sum_i v_{xi}^2 \, dh_i$$

$$\beta_m \cong \frac{1}{3.7 \, ft}$$

$$\times \frac{s^2}{(0.85)^2 \, ft^2} [0.55^2 + 0.85^2 + 1.0^2 + (1.1^2 \times 0.7)] \frac{ft^3}{s^2} = 1.074$$

(k) Energy correction factor [Eq. (E3.6.2)]:

$$\alpha_e = \frac{1}{AV_x^3} \int_A v_x^3 \, dA \cong \frac{1}{hV_x^3} \sum_i v_{xi}^3 \, dh_i$$

$$\alpha_e \cong \frac{1}{3.7 \, ft}$$

$$\times \frac{s^3}{(0.85)^3 \, ft^3} [0.55^3 + 0.85^3 + 1.0^3 + (1.1^3 \times 0.7)] \frac{ft^4}{s^3} = 1.194$$

6.4　Deviation from logarithmic velocity profiles

Two types of deviation to the logarithmic velocity profiles are considered: beyond the law of the wall (Section 6.4.1) and in narrow channels (Section 6.4.2).

6.4.1　*Wake flow function*

Departure from logarithmic velocity profiles is observed as the distance from the boundary increases (see dashed line $u_* z/\nu_m > 1{,}000$ in Fig. 6.4). The reason for this is essentially related to the invalidity of the following assumptions: (1) constant shear stress throughout the fluid and (2) mixing length approximation $l_m = \kappa z$.

A more complete description of the velocity distribution v_x, including the law of the wake for steady turbulent open-channel flow, has been suggested by Coles (1956):

$$\frac{v_x}{u_*} = \underbrace{\left[\frac{2.3}{\kappa}\log\left(\frac{u_* z}{\nu_m}\right)+5.5\right]}_{\text{law of the wall}} - \underbrace{\frac{\Delta v_x}{u_*}}_{\substack{\text{roughness}\\\text{function}}} + \underbrace{\frac{2\Pi_w}{\kappa}\sin^2\left(\frac{\pi z}{2h}\right)}_{\text{wake flow function}} \qquad (6.24)$$

where h is the total flow depth.

The terms in brackets depict the original logarithmic law of the wall for smooth boundaries from Equation (6.20). The last two terms have been added to describe the entire boundary layer velocity profile outside of the thin laminar sublayer. The term $\Delta v_x/u_*$ is the channel roughness velocity reduction function. The last term describes the velocity increase in the wake region as described by the wake strength coefficient Π_w.

The wake flow function equals zero near the boundary and increases gradually toward $2\Pi_w/\kappa$ at the upper surface ($z = h$). With $v_x = v_{xm}$ at $z = h$, the upper limit of the velocity profile is

$$\frac{v_{xm}}{u_*} = \frac{2.3}{\kappa}\log\left(\frac{u_* h}{\nu_m}\right)+5.5-\frac{\Delta v_x}{u_*}+\frac{2\Pi_w}{\kappa} \qquad (6.25)$$

The velocity defect law obtained after subtracting Equation (6.24) from Equation (6.25) gives

$$\frac{v_{xm}-v_x}{u_*} = \left\{\frac{2\Pi_w}{\kappa} - \left[\frac{2.3}{\kappa}\log\frac{z}{h}\right]\right\} - \frac{2\Pi_w}{\kappa}\sin^2\left(\frac{\pi z}{2h}\right) \qquad (6.26) \blacklozenge$$

In this form, the term in brackets is the original velocity defect equation for the logarithmic law. The wake flow term vanishes as z approaches zero, and the velocity defect asymptotically reaches the term in braces in Equation (6.26) as z/h diminishes. This means that the von Kármán constant κ must be defined from the slope of the logarithmic part in the lower portion of the velocity profile. The wake strength coefficient Π_w is then determined by projecting a straight line, fit in the lower portion of the velocity profile, to $z/h = 1$ and calculating Π_w from

$$\Pi_w = \frac{\kappa}{2}\left[\frac{v_{xm}-v_x}{u_*}\right] \quad \text{at } z/h = 1 \qquad (6.27)$$

This procedure, illustrated in Figure 6.5, generally shows that Π_w increases with sediment concentration, while the von Kármán κ remains constant around 0.4.

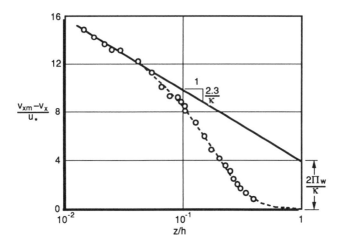

Figure 6.5. Evaluation of κ and Π_w from the velocity defect law

6.4.2   Sidewall correction method

Consider steady uniform flow in a narrow open channel at a discharge Q measured from a calibrated orifice and friction slope S_f. In smooth-walled laboratory flumes where the flume width W is less than five times the flow depth h, the sidewall resistance is different from the bed resistance. The Vanoni–Brooks correction method can be applied to determine the bed shear stress $\tau_b = \rho_m u_{*b}^2$. For a rectangular channel, the hydraulic radius $R_h = Wh/W + 2h$ and the Reynolds number $\mathrm{Re} = 4VR_h/\nu_m$ are calculated given the average velocity $V = Q/Wh$. The shear velocity $u_* = \sqrt{gR_hS_f}$ computed from the slope S_f is then used to calculate the Darcy–Weisbach friction factor $f = 8u_*^2/V^2$. The wall friction factor f_w for turbulent flow over a smooth boundary, $10^5 < \mathrm{Re}/f < 10^8$, can be calculated from

$$f_w = 0.0026\left\{\log\left(\frac{\mathrm{Re}}{f}\right)\right\}^2 - 0.0428\log\left(\frac{\mathrm{Re}}{f}\right) + 0.1884 \qquad (6.28a)$$

The bed friction factor f_b is then obtained from

$$f_b = f + \frac{2h}{W}(f - f_w) \qquad (6.28b)$$

The hydraulic radius related to the bed $R_b = R_h f_b/f$ is then used to calculate the bed shear stress τ_b from $\tau_b = \gamma_m R_b S_f$. Example 6.2 provides the details of the calculation procedure.

Example 6.2 Application of the sidewall correction method. Consider a discharge $Q = 1.2$ ft^3/s in a 4-ft-wide flume inclined at a 0.001 slope. Calculate the bed shear stress τ_b given the normal flow depth of 0.27 ft. The measured water temperature is 70°F.

Step 1

$$Q = 1.2 \text{ ft}^3/\text{s}, \qquad S_0 = S_f = 0.001, \qquad h_n = 0.27 \text{ ft}, \qquad \nu_m \cong 1 \times 10^{-5} \text{ ft}^2/\text{s}$$

Step 2. Hydraulic radius:

$$R_h = \frac{W h_n}{W + 2 h_n} = \frac{4 \text{ ft} \times 0.27 \text{ ft}}{(4 + 2 \times 0.27) \text{ ft}} = 0.238 \text{ ft}$$

Step 3. Flow velocity:

$$V = \frac{Q}{W h_n} = \frac{1.2 \text{ ft}^3}{\text{s} \times 4 \text{ ft} \times 0.27 \text{ ft}} = 1.11 \frac{\text{ft}}{\text{s}}$$

Step 4. Reynolds number:

$$\text{Re} = \frac{4 V R_h}{\nu_m} = 1.06 \times 10^5$$

Step 5. Shear velocity:

$$u_* = \sqrt{g R_h S_f} = \sqrt{32.2 \text{ ft/s}^2 \times 0.238 \text{ ft} \times 0.001} = 0.087 \text{ ft/s}$$

Step 6. Darcy–Weisbach factor:

$$f = \frac{8 u_*^2}{V^2} = \frac{8(0.087)^2}{(1.11)^2} \frac{\text{ft}^2}{\text{ft}^2} \frac{\text{s}^2}{\text{s}^2} = 0.049$$

Step 7. Wall friction factor:

$$f_w = 0.0026 \left[\log \frac{\text{Re}}{f} \right]^2 - 0.0428 \left[\log \frac{\text{Re}}{f} \right] + 0.1884$$

$$f_w = 0.021$$

Step 8. Bed friction factor:

$$f_b = f + \frac{2 h_n}{W} (f - f_w)$$

$$f_b = 0.049 + \frac{2 \times 0.27 \text{ ft}}{4 \text{ ft}} (0.049 - 0.021) = 0.0527$$

Step 9. Bed hydraulic radius:

$$R_b = \frac{f_b}{f} R_h = \frac{0.0527}{0.049} \times 0.238 \text{ ft} = 0.255 \text{ ft}$$

Step 10. Bed shear stress:

$$\tau_b = \gamma_m R_b S_f = \frac{62.3 \text{ lb}}{\text{ft}^3} \times 0.255 \text{ ft} \times 0.001 = 0.016 \frac{\text{lb}}{\text{ft}^2}$$

6.5   Open-channel flow measurements

Open-channel flow measurements normally include stage and flow velocity measurements.

6.5.1   *Stage measurements*

Stage measurements determine water surface elevation with reference to a datum such as the mean sea level, a local datum related to project activity, a reference elevation plane, a benchmark, or an arbitrary datum below the elevation of zero flow. Simple nonrecording gages require frequent readings to develop continuous water level records. Nonrecording gages are either directly read or provide measurements of the water surface elevation at a fixed point.

Staff gages are usually vertical boards or rods precisely graduated with reference to a datum.

Point gages consist of mechanical devices to locate and measure the water surface elevation. Measurements can be taken from a graduated rod, drum, or steel tape housed in a small box mounted on a rigid structure (such as a bridge) directly above the water surface.

Float gages are used primarily with an analog water stage recorder. The gage consists of a float and counterweight connected by a graduated steel tape, which passes over a pulley assembly. A relatively large float and counterweight are required for stability, sensitivity, and accuracy – such as a 10-in. copper float and a 2-lb lead counterweight.

Pressure-type gages use water pressure transmitted through a tube to a manometer inside a gage shelter to measure stage. Stage can also be measured by gas bubbling freely into a stream from a submerged tube set at a fixed elevation; the gage pressure in the tube equals the piezometric head at the open end of the tube.

Crest-stage gages measure maximum flood stage from granulated cork floats as the water rises in the pipe. When the water recedes, the cork adheres to the pipe, marking the crest stage.

A recording stage gage produces a punched, printed, traced, analog or digital record of water surface elevation with respect to time. The gage height recording is usually activated by either a float mechanism or a pressure-sensing device. The strip-chart records show an uninterrupted recording of water-level fluctuations with time. Digital recorders store, punch, or print out gage heights at preselected time intervals.

Analog recorders provide a continuous visual record of stage useful for graphical presentation.

Digital recorders provide data in digitally coded form suitable for digital computer processing. Because digital systems record gage height only at preselected time intervals, maximum and minimum peak gage heights of a flashy stream cannot be accurately measured.

Telemetering systems using telephone, radio, or satellite communication are desirable when current information on stage is frequently needed from remote locations. Some telemetering sysems continuously indicate or record stage at a given site; others report instantaneous gage readings on request.

6.5.2 Velocity measurements

Velocity is measured with such devices as floats, drag bodies, tracers, rotating-element current meters, and deflection vanes, and by means of optical, laser, ultrasonic, and velocity-head methods.

Rotating current meters are based on the proportionality between the angular velocity of the rotation device and the flow velocity. By counting the number of revolutions of the rotor in a measured time interval, point velocity is determined. A vertical-axis current meter measures the differential drag on two sides of cups in relative motion in a fluid. The rotation speed of these devices, such as Price current meters, is calibrated against the fluid velocity. Horizontal-axis current meters act as propellers in a moving fluid. Common horizontal-axis meters include the Ott and the Neyrpic current meters.

Mechanical meters are limited in both high and low velocities; electromagnetic meters are less so. They are easier to use, have no moving parts, are generally more accurate, indicate both velocity and direction, and provide electronic readout with averaging.

Acoustic (ultrasonic) velocity meters measure velocity by determining the travel time of sound pulses moving in both directions along a diagonal path between transducers mounted near each bank of the stream. The water velocity is the average velocity component parallel to the acoustic path line between the two transducers. Acoustic velocity meter systems operate satisfactorily in the laboratory, but they are usually complex and expensive.

Ultrasonic Doppler velocimeters measure the phase shift between the signal emitted along the upstream path and the scattered signal received in the opposite direction.

Hot film and hot wire anemometers are electrically heated sensors being cooled by advection. The heat loss being a function of the flow velocity, this laboratory instrument is calibrated to measure fluctuating velocities with high spatial resolution and high-frequency response.

Laser Doppler anemometers measure the Doppler frequency shift of light-scattering particles moving with the fluid. The frequency shift provides very accurate flow velocity measurements in the laboratory without flow disturbance.

Electromagnetic flow meters are based on Faraday's induction law stating that voltage is induced by the motion of a conductor (fluid) perpendicular to a magnetic field.

The depth-averaged velocity is normally obtained from a measured velocity profile. The following approximate methods for turbulent flows can be used to determine the depth-averaged flow velocity from point velocity measurements at one, two, or three points:

1. The one-point method (at 60% of the total depth measured down from the water surface) uses the observed velocity at 0.6 h as the mean velocity in the vertical. It gives reliable results in uniform cross sections without large irregularities.

2. The two-point method (at 20% and 80% of the total depth measured down from the water surface) averages the velocities at the two depths, and the average fairly approximates the mean velocity along the vertical.

3. The three point method (at 20%, 60%, and 80% of the total depth measured down from the water surface) combines the one-point and two-point methods. Velocities at 0.2 and 0.8 h are averaged, and the value is then averaged with the 0.6 depth velocity measurement to obtain fairly accurate values of depth-averaged flow velocity.

4. The surface method, which has only limited use, assumes a coefficient (usually about 0.85) to convert the surface velocity measured with a float to the depth-averaged velocity. This method is not very accurate.

Exercises

*6.1. Substitute Equations (6.1a–d) into the Navier–Stokes equations (Table 5.1) to obtain Equation (6.4a).

*6.2. Demonstrate that Equation (6.13) is obtained from Equation (6.12) when $z_0 = k_s'/30$.

*6.3. Demonstrate that Equation (6.20) is obtained from Equation (6.12) when $z_0 = \nu_m/9u_*$.

*6.4. Derive Equation (6.23) from Equations (6.22b) and (6.20) at $z = \delta$.

**Problem 6.1

Consider the clear-water and sediment-laden velocity profiles measured in a smooth laboratory flume at a constant discharge by Coleman (1986). Notice the changes in the velocity profiles due to the presence of sediments. Determine the von Kármán constant κ from Equation (6.12) for the two velocity profiles in the following tabulation, given $u_* = 0.041$ m/s, $d_s = 0.105$ mm, $Q = 0.064$ m^3/s, $h \cong 0.17$ m, $S_f = 0.002$, and $W = 0.356$ m.

Elevation[a] (mm)	Clear-water flow velocity (m/s)	Sediment-laden velocity (m/s)	Concentration by volume
6	0.709	0.576	2.1×10^{-2}
12	0.773	0.649	1.2×10^{-2}
18	0.823	0.743	7.7×10^{-3}
24	0.849	0.798	5.9×10^{-3}
30	0.884	0.838	4.8×10^{-3}
46	0.927	0.916	3.2×10^{-3}
69	0.981	0.976	2.5×10^{-3}
91	1.026	1.047	1.6×10^{-3}
122	1.054	1.07	8.0×10^{-4}
137	1.053	1.07	—
152	1.048	1.057	—
162	1.039	1.048	—

[a] Elevation above the bed.

Answer: The von Kármán constant κ remains close to 0.4 when the lowest portion of both velocity profiles is considered. When the main portion of the velocity profiles is considered, κ becomes significantly smaller for sediment-laden flow.

*Problem 6.2

(a) In turbulent flows, determine the elevation at which the local velocity v_x is equal to the depth-averaged velocity V_x. (*Hint:* $V_x = 1/(h-z_0)\int_{z_0}^{h} v_x \, dz$.)
(b) Determine the elevation at which the local velocity v_x equals the shear velocity u_*.

**Problem 6.3

Determine the Darcy–Weisbach friction factor f from the data in Problem 6.1.

Answer

$$f = \frac{8u_*^2}{V^2} = 0.012$$

**Problem 6.4

(a) Calculate the laminar sublayer thickness δ in Problem 6.1.
(b) Estimate the range of laminar sublayer thicknesses for bed slopes $1 \times 10^{-5} < S_0 < 0.01$ and flow depths $0.5 \text{ m} < h < 5 \text{ m}$.

**Problem 6.5

From turbulent velocity measurements at two elevations (v_1 at z_1 and v_2 at z_2) in a wide rectangular channel, use Equation (6.12) to determine the shear velocity u_*; the boundary shear stress τ_0; and the laminar sublayer thickness δ.

Answer

$$u_* = \frac{\kappa(v_1 - v_2)}{\ln(z_1/z_2)}$$

$$\tau_0 = \rho_m u_*^2 = \rho_m \kappa^2 \left[\frac{v_1 - v_2}{\ln(z_1/z_2)}\right]^2$$

$$\delta = \frac{11.6\nu_m}{u_*} = 11.6\frac{\nu_m \ln(z_1/z_2)}{\kappa(v_1 - v_2)}$$

*Problem 6.6

(a) With reference to Problem 6.1, evaluate the parameters κ and Π_w from the velocity defect formulation in Equation (6.26). Compare the value of κ with the value obtained previously (Problem 6.1) from Equation (6.12).

(b) Compare the experimental velocity profiles from Problem 6.1 with the velocity profiles calculated from Equations (6.13) and (6.20).

**Problem 6.7

For the velocity profile given in Problem 6.1, calculate the depth-averaged velocity from (a) the velocity profile; (b) the one-point method; (c) the two-point method; (d) the three-point method; and (e) the surface method.

7

Incipient motion

The threshold conditions between erosion and sedimentation of a single particle are usually referred to as incipient motion. The stability of granular material is first examined without flow in Section 7.1 and under flowing water in the following sections. In Section 7.2, simplified particle equilibrium conditions on nearly horizontal surfaces are discussed for uniform and nonuniform material. The equilibrium of particles under tridimensional moments of force is treated in Section 7.3. A simplified force analysis is presented in Section 7.4. Three examples of particle stability analysis and stable channel design are provided.

7.1   Angle of repose

The stability of a single particle on a plane horizontal surface is first considered in Figure 7.1a for simple two-dimensional particle shapes. The threshold condition is obtained when the particle center of mass is vertically above the point of contact. It is shown that the critical angle at

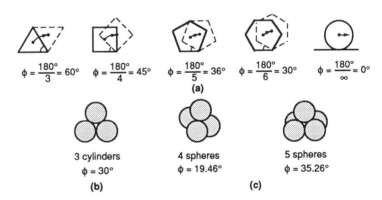

$$\phi = \frac{180°}{3} = 60° \quad \phi = \frac{180°}{4} = 45° \quad \phi = \frac{180°}{5} = 36° \quad \phi = \frac{180°}{6} = 30° \quad \phi = \frac{180°}{\infty} = 0°$$

(a)

3 cylinders
$\phi = 30°$
(b)

4 spheres
$\phi = 19.46°$

5 spheres
$\phi = 35.26°$

(c)

Figure 7.1. Angle of repose for simple shapes

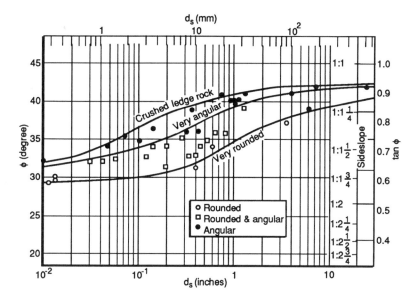

Figure 7.2. Angle of repose for granular material
(from Simons, 1957)

which motion occurs, or angle of repose ϕ, equals 180° divided by the
number of sides of the polygon. For instance, the angle of repose ϕ of an
equilateral triangle is $\phi = 180°/3 = 60°$, that of a square is $\phi = 180°/4 =$
45°, and that of a sphere is $\phi = 180°/\infty = 0°$. One concludes that the angle
of repose of particles on a flat surface increases with angularity.

Long cylinders standing on each other (Fig. 7.1b) rest at an angle of repose $\phi = 30°$. A sphere standing on spheres of equal diameter (Fig. 7.1c)
reaches threshold of motion at $\phi = 19.5°$ for three points of contact, and
at $\phi = 35.3°$ for four points of contact. For natural granular material,
the angle of repose varies with grain size and angularity of the material,
as shown in Figure 7.2.

In the case of material with different diameter particles, consider a
sphere of diameter d_2 resting on top of four identical spheres of different
diameter d_1, as sketched in Figure 7.3. Geometrically, the angle of repose
is given by

$$\tan \phi = \frac{d_1}{\sqrt{(d_1+d_2)^2 - 2d_1^2}} = \sqrt{\frac{1}{(1+d_2/d_1)^2 - 2}} \tag{7.1}$$

The angle of repose ϕ is 35.3° when $d_2 = d_1$; it decreases when $d_2 > d_1$
and increases for fine particles $d_2 < d_1$ until $d_2 = 0.41d_1$.

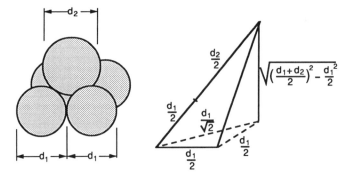

Figure 7.3. Angle of repose for spheres of different diameter

7.2   Submerged incipient motion

Fluid flow around sediment particles exerts forces that tend to initiate particle motion. The resisting force of noncohesive material relates to particle weight. Threshold conditions occur when the hydrodynamic moment of forces acting on a single particle balances the resisting moment of force. The particle is then at incipient motion.

The forces acting on a noncohesive sediment particle sketched in Figure 7.4 are the particle weight F_W, buoyancy force F_B, lift force F_L, drag force F_D, and resisting force F_R.

From the analysis of the hydrodynamic forces around a sphere in Example 5.1, no lift forces could be generated under creeping flow conditions. In turbulent flows, however, lift forces depend on the circulation around the particle. As a first approximation, it is assumed that the lift and drag forces are proportional; a refined analysis is presented in Section 7.3. Further assuming that the bed surface slope is very flat, $S_0 = \tan \Theta \cong 0$, and that the water surface is almost horizontal, the buoyancy force F_B acts in a vertical direction, opposite to the particle weight F_W. The submerged weight of the particle, $F_S = F_W - F_B$, is therefore considered. Assuming that the moment arms are equivalent, the ratio of hydrodynamic forces $F_L \sim F_D \sim \tau_0 d_s^2$ to the submerged weight $F_S \sim (\gamma_s - \gamma_m) d_s^3$ defines the dimensionless shear stress τ_*, called the Shields parameter:

$$\tau_* = \frac{\tau_0}{(\gamma_s - \gamma_m)d_s} = \frac{\rho_m u_*^2}{(\gamma_s - \gamma_m)d_s} \tag{7.2}$$

where

$\tau_0 =$ boundary shear stress

$u_* =$ shear velocity

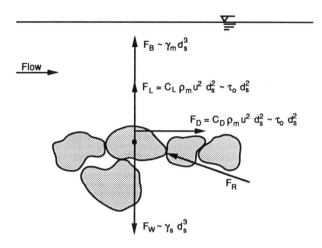

Figure 7.4. Force diagram under steady uniform flow

γ_s = specific weight of a sediment particle
γ_m = specific weight of the fluid mixture
d_s = particle size

7.2.1 Uniform grain size

The critical value of the Shields parameter τ_{*c} corresponding to the beginning of motion ($\tau_0 = \tau_c$) depends on whether laminar or turbulent flow conditions prevail around the particle. Besides the angle of repose, one should consider the ratio of sediment size to the laminar sublayer thickness expressed either as d_s/δ or as the grain shear Reynolds number $Re_* = u_* d_s/\nu_m$ (because $\delta = 11.6\nu_m/u_*$):

$$\tau_{*c} = \frac{\tau_c}{(\gamma_s - \gamma_m)d_s} = \mathcal{F}\left(\frac{u_* d_s}{\nu_m}, \tan\phi\right) \tag{7.3a}$$

$$\tau_{*c} \cong 0.06 \tan\phi \quad \text{when} \quad \frac{u_* d_s}{\nu_m} > 50 \tag{7.3b}$$

Shields (1936) determined the threshold condition by measuring sediment transport for values of τ_* at least twice as large as the critical value and then extrapolated to the point of vanishing sediment transport. His laboratory experiments and those of Yalin and Karahan (1979) using the median grain size for d_s led to the modified Shields diagram shown in Figure 7.5.

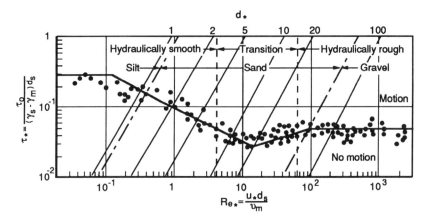

Figure 7.5. Threshold of motion for granular material

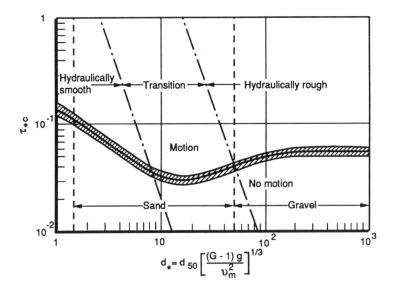

Figure 7.6. Particle motion diagram

Since the shear velocity u_* appears both in the Shields parameter $\tau_* = \rho_m u_*^2/(\gamma_s - \gamma_m)d_s$ and in the grain shear Reynolds number $\mathrm{Re}_* = u_* d_s/\nu_m$, it is possible to replace the abscissa of the Shields diagram after eliminating the shear velocity from Re_* and defining the dimensionless particle diameter d_* from $d_*^3 = \mathrm{Re}_*^2/\tau_*$. Thus, the abscissa of the Shields diagram can be replaced by the dimensionless particle diameter, resulting in Figure 7.6.

The critical values of the Shields parameter τ_{*c} can be approximated as follows:

$$\tau_{*c} = 0.5 \tan \phi \qquad \text{when } d_* < 0.3 \tag{7.4a}$$

$$\tau_{*c} = 0.25 d_*^{-0.6} \tan \phi \quad \text{when } 0.3 < d_* < 19 \tag{7.4b}$$

$$\tau_{*c} = 0.013 d_*^{0.4} \tan \phi \quad \text{when } 19 < d_* < 50 \tag{7.4c}$$

$$\tau_{*c} = 0.06 \tan \phi \qquad \text{when } d_* > 50 \tag{7.4d}$$

where

$$d_* = d_s \left[\frac{(G-1)g}{\nu_m^2} \right]^{1/3} \tag{7.4e}$$

For turbulent flows over rough boundaries, the critical shear stress becomes linearly proportional to the sediment size, since the threshold value of the Shields parameter remains constant. Based on a comparison of data from the Highway Research Board (1970), a graphical relationship between critical shear stress τ_c and median grain size d_{50} on a flat horizontal surface is shown in Figure 7.7 with equivalent values for granular

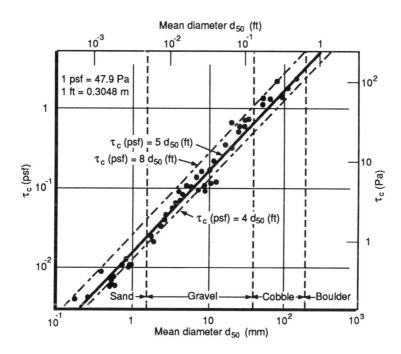

Figure 7.7. Critical shear stress on a horizontal surface

Table 7.1. *Approximate threshold conditions for granular material at 20° C*

Class name	d_s (mm)	d_*	ϕ (deg)	τ_{*c}	τ_c (Pa)	u_{*c} (m/s)
Boulder						
Very large	>2,048	51,800	42	0.054	1,790	1.33
Large	>1,024	25,900	42	0.054	895	0.94
Medium	>512	12,950	42	0.054	447	0.67
Small	>256	6,475	42	0.054	223	0.47
Cobble						
Large	>128	3,235	42	0.054	111	0.33
Small	>64	1,620	41	0.052	53	0.23
Gravel						
Very coarse	>32	810	40	0.05	26	0.16
Coarse	>16	404	38	0.047	12	0.11
Medium	>8	202	36	0.044	5.7	0.074
Fine	>4	101	35	0.042	2.71	0.052
Very fine	>2	50	33	0.039	1.26	0.036
Sand						
Very coarse	>1	25	32	0.029	0.47	0.0216
Coarse	>0.5	12.5	31	0.033	0.27	0.0164
Medium	>0.25	6.3	30	0.048	0.194	0.0139
Fine	>0.125	3.2	30	0.072	0.145	0.0120
Very fine	>0.0625	1.6	30	0.109	0.110	0.0105
Silt						
Coarse	>0.031	0.8	30	0.165	0.083	0.0091
Medium	>0.016	0.4	30	0.25	0.065	0.0080

material in Table 7.1. Values of the maximum permissible velocity V_c in canals from Fortier and Scobey (1926) and Mirtskhoulava (1988) are listed in Table 7.2 for mostly noncohesive material and in Table 7.3 for cohesive material.

7.2.2   Nonuniform grain size

The particle size distribution of bed material depends on the magnitude of the applied shear stress in relation to the mobility of sediment particles of different sizes. Consider a sediment mixture, identical for the three cases illustrated in Figure 7.8, for which the applied shear stress can move: (a) none of the particles, (b) the fine particles only, and (c) all the particles.

Table 7.2. *Maximum permissible velocities for canals with* h < 1 m

Soil type	Manning coefficient n	Clear water, no detritus (m/s)	Water transporting colloidal silt (m/s)	Water transporting noncolloidal silts, sands, gravels, or rock fragments (m/s)
Stiff clay (very colloidal)	.025	1.14	1.52	0.91
Alluvial silt when colloidal	.025	1.14	1.52	0.91
Alluvial silt when noncolloidal	.02	0.61	1.07	0.61
Volcanic ash	.02	0.76	1.07	0.61
Silt loam (noncolloidal)	.02	0.61	0.91	0.61
Ordinary firm loam	.02	0.76	1.07	0.69
Sandy loam (noncolloidal)	.02	0.53	0.76	0.61
Fine sand (colloidal)	.02	0.46	0.76	0.46
Fine gravel	.02	0.76	1.52	1.14
Graded, loam to cobbles, when noncolloidal	.03	1.14	1.52	1.52
Graded, silt to cobbles, when colloidal	.03	1.22	1.68	1.52
Coarse gravel (noncolloidal)	.025	1.22	1.83	1.98
Cobbles and shingles	.035	1.52	1.68	1.98
Shales and hard pans	.025	1.83	1.83	1.52

Source: Fortier and Scobey (1926).

It is observed that, without motion of the fines, the bed surface is composed of the original bed material.

With motion of the finer fractions at a shear stress insufficiently high to displace the coarse particles, the bed surface is coarsened to form an armor layer while the fractions in motion are finer than those of the bed surface. Finally, as the shear stress becomes sufficiently large to break the coarse armor layer, the finer particles of the moving layer are displaced to the surface of the bed, thus creating a smooth surface on top of which the coarse particles easily roll. This type of motion is conducive to lamination and stratification (Julien, Lan, and Berthault 1993). In this case, it is observed that the size fractions in motion in the channel can be coarser than the fractions in the bed. One thus notices that, starting from the same bed-material size distribution, the surface layer of the bed can be coarser than d_{50} of the original mixture (Fig. 7.8b), whereas the surface layer can become finer than d_{50} of the original mixture (Fig. 7.8c),

Table 7.3. *Summary of maximum permissible average flow velocities for cohesive channels*

Soil type	Maximum velocity (m/s)
Fine sandy loamy clay	0.45–0.91
Alluvial mud	0.61–0.84
Alluvial loamy clay	0.76–0.84
Hard loamy clay	0.91–1.14
Hard clay	0.76–1.52
Rigid clay	1.22–1.52
Clayey shale	0.76–2.13
Hard rock	3–4.5

ρ_m (kg/m³)	Loamy sand[a] (m/s)	Non-plastic clay (m/s)	Clay[a] (m/s)	Heavy clayey soil (m/s)	Loamy clay[a] (m/s)
1,200	$0.1 \log 8.8h/k_s$	—	$0.12 \log 8.8h/k_s$	—	$0.12 \log 8.8h/k_s$
1,200–1,650	$0.15 \log 8.8h/k_s$	—	$0.3 \log 8.8h/k_s$	—	$0.25 \log 8.8h/k_s$
1,544	—	0.32	0.35	0.4	0.45
1,650–2,040	—	—	$0.45 \log 8.8h/k_s$	—	$0.4 \log 8.8h/k_s$
1,742	—	0.7	0.8	0.85	0.9
2,040	—	1.05	1.2	1.25	1.3
2,040–2,140	—	—	$0.65 \log 8.8h/k_s$	—	$0.6 \log 8.8h/k_s$
2,270	—	1.35	1.65	1.70	1.8

[a] h is the flow depth and k_s the boundary roughness height.
Source: Modified after Etcheverry (1916), Fortier and Scobey (1926), and Mirtskhoulava (1988).

depending on the relative magnitude of the applied shear stress to the critical shear stress by size fractions.

The analysis of armor layers (Fig. 7.8b) centers around the coarser fractions of the mixture that are not moving; finer fractions will be present in the mixture as long as they are shielded by the stable coarser particles. All particles will enter motion as soon as the shear stress exceeds the threshold of motion of coarse particles. This concept is referred to as the equal-mobility concept for which all fractions of sediment enter motion at the same value of applied shear stress.

In the case of graded sediment mixtures, the beginning of motion of a noncohesive sediment particle of size d_i in a mixture of average grain size d_{50} is given by the critical Shields parameter $\tau_{*ci} = \tau_{ci}/(\gamma_s - \gamma_m)d_i$ for fraction i with reference to the critical dimensionless shear stress $\tau_{*c50} = \tau_{c50}/(\gamma_s - \gamma_m)d_{50} \cong 0.083$ at which the grain size d_{50} enters motion. The

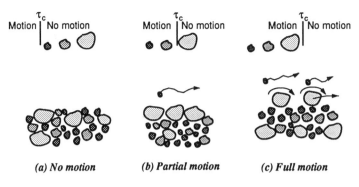

(a) No motion   *(b) Partial motion*   *(c) Full motion*

Figure 7.8. Bed surface for sediment mixtures

graphical relationship between τ_{*ci}/τ_{*c50} and the grain size ratio d_i/d_{50} is shown in Figure 7.9 with experimental data compiled by Tsujimoto (1992) and Garde and Ranga Raju (1985). The equal-mobility concept is described by $\tau_{*ci}d_i = \tau_{*c50}d_{50}$.

The analysis of graded mixtures under full motion (Fig. 7.8c) deserves appropriate consideration of the angle of repose from Equation (7.1). The critical shear stress τ_{c21} for a particle of diameter d_2 resting on a bed of particles of diameter d_1 can be compared with the critical shear stress τ_{c1} for a particle of diameter d_1 resting on particles of the same size d_1 from Equation (7.1):

$$\frac{\tau_{c21}}{\tau_{c1}} = \frac{d_2}{d_1}\sqrt{\frac{2}{(1+d_2/d_1)^2-2}} \qquad (7.5)$$

This relationship, plotted with $d_2 = d_i$ and $d_1 = d_{50}$ in Figure 7.9, shows that the particles of diameter $d_2 > d_1$ are almost as mobile as particles of the same size, $d_2 = d_1$.

7.3   Moment stability analysis

Layers of large stones, commonly called riprap, are used to protect embankment slopes against erosion. The stability of riprap depends on the stability of individual particles subjected to hydrodynamic forces under various embankment configurations and stone properties.

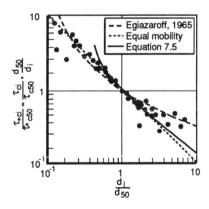

Figure 7.9. Critical shear stress by size fraction

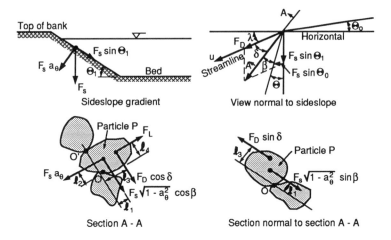

Figure 7.10. Moment stability analysis of a particle

Figure 7.10 illustrates the forces acting on a cohesionless particle resting on an embankment inclined at a sideslope angle Θ_1 and a downstream angle Θ_0. These are the lift force F_L, the drag force F_D, the buoyancy force F_B, and the weight of the particle F_W. As long as the water surface slope angle in the downstream direction is small, the buoyancy force can be subtracted from the weight of the particle to give the submerged weight of the particle, $F_S = F_W - F_B$. The lift force is defined as the fluid force normal to the embankment plane, whereas the drag force is acting along the plane in the same direction as the velocity field surrounding the particle.

Geometrically, one defines

$$a_\Theta = \sqrt{\cos^2\Theta_1 - \sin^2\Theta_0} \quad \text{and} \quad \tan\Theta = \frac{\sin\Theta_0}{\sin\Theta_1}$$

The submerged weight has one sideslope component, $F_S \sin\Theta_1$; one downslope component, $F_S \sin\Theta_0$; and a component normal to the plane, $F_S a_\Theta$. The streamline deviates from the horizontal at an angle λ along the embankment plane (λ is defined positive downward). Once in motion, the particle follows a direction at an angle β from the downward direction (projection of a vertical on an embankment plane).

Stability against rotation of a particle determines incipient conditions of motion when the equilibrium of moments about the point of rotation 0 is satisfied (see Fig. 7.10, Section A–A):

$$l_2 F_S a_\Theta = l_1 F_S \sqrt{1 - a_\Theta^2} \cos\beta + l_3 F_D \cos\delta + l_4 F_L \tag{7.6}$$

The angles δ and β and the moment arms l_1, l_2, l_3, and l_4 are shown in Figure 7.10.

Two stability factors, SF_0 and SF_{01}, against rotation about points 0 and $0'$ are respectively defined as the ratio of the resisting moments to the moments generating motion:

$$SF_0 = \frac{l_2 F_S a_\Theta}{l_1 F_S \sqrt{1 - a_\Theta^2} \cos\beta + l_3 F_D \cos\delta + l_4 F_L} \tag{7.7a}$$

$$SF_{01} = \frac{l_2 F_S a_\Theta + l_1 F_S \sqrt{1 - a_\Theta^2} \cos\beta}{l_3 F_D \cos\delta + l_4 F_L} \tag{7.7b}$$

Because the stability factor SF_0 equals unity when the angle Θ equals the angle of repose ϕ under static fluid conditions ($F_D = F_L = 0$), it is found that $\tan\phi = l_2/l_1$. Dividing both the numerator and the denominator by $l_1 F_S$ transforms Equation (7.7) to

$$SF_0 = \frac{a_\Theta \tan\phi}{\eta_1 \tan\phi + \sqrt{1 - a_\Theta^2} \cos\beta} \tag{7.8a} \blacklozenge$$

$$SF_{01} = \frac{a_\Theta \tan\phi + \sqrt{1 - a_\Theta^2} \cos\beta}{\eta_1 \tan\phi} \tag{7.8b}$$

$$\eta_1 = M + N\cos\delta \tag{7.9}$$

where $M = (l_4/l_2)(F_L/F_S)$ and $N = (l_3/l_2)(F_D/F_S)$. The variable η_1 is called the stability number for the particles on the embankment side-slope. The variable η_1 relates to the stability number $\eta_0 = M + N$ for particles on a plane horizontal surface ($\Theta_0 = \Theta_1 = \delta = 0$) after considering $\lambda + \delta + \beta + \Theta = 90°$:

$$\eta_1 = \eta_0 \left\{ \frac{(M/N) + \sin(\lambda + \beta + \Theta)}{1 + (M/N)} \right\} \tag{7.10} \blacklozenge$$

When the flow is fully turbulent over a hydraulically rough horizontal surface, incipient motion corresponds to $SF_0 = 1$, or

$$\eta_0 = \frac{\tau_0}{\tau_c} \cong \frac{21\tau_0}{(\gamma_s - \gamma_m)d_s} \tag{7.11a} \blacklozenge$$

This normalized form of the Shields parameter shows that $\eta_0 = 1$ when $\tau_{*c} = 0.047$, describing the incipient motion of particles on a plane bed under turbulent flow over hydraulically rough boundaries.

Alternative relationships for η_0 are obtained when the boundary shear stress τ_0 is replaced by the reference velocity v_r, the velocity against the particle v_p, or the average flow velocity V. The reference velocity v_r is the velocity at a height $z = k'_s$; thus, from Equation (6.13), $\tau_0 = \rho_m u_*^2 = 0.0138\rho_m v_r^2$. Taking the velocity against the particle $v_p \cong 0.71 v_r$, we have $\tau_0 = 0.027\rho_m v_p^2$. Finally, from Equation (6.14),

$$\tau_0 = \tau'_0 = \rho_m u_*'^2 = f' \frac{\rho_m V^2}{8} = \frac{\rho_m V^2}{[5.75 \log(12.2h/k'_s)]^2}$$

and alternative equations for η_0 as a function of v_r, v_p, and V are, respectively,

$$\eta_0 = \frac{0.3 v_r^2}{(G-1)gd_s} \tag{7.11b}$$

$$\eta_0 = \frac{0.6 v_p^2}{(G-1)gd_s} \tag{7.11c}$$

$$\eta_0 = \frac{V^2}{(G-1)gd_s[5.75 \log(12.2h/k'_s)]^2} \tag{7.11d}$$

The second equilibrium condition given by the direction of the particle along the section normal to Section A–A in Figure 7.10 is

$$l_3 F_D \sin \delta = l_1 F_S \sqrt{1 - a_\Theta^2} \sin \beta \tag{7.12}$$

Writing δ as a function of λ, Θ, and β and solving for β gives

$$\beta = \tan^{-1}\left\{ \cos(\lambda + \Theta) \middle/ \left[\frac{(M+N)\sqrt{1 - a_\Theta^2}}{N\eta_0 \tan \phi} + \sin(\lambda + \Theta) \right] \right\} \tag{7.13} \blacklozenge$$

In summary, the stability factors for particles on sideslopes can be calculated from Equation (7.8) after solving successively Equations (7.11), (7.13), and (7.10), with the use of two geometric relationships,

$$a_\Theta = \sqrt{\cos^2 \Theta_1 - \sin^2 \Theta_0} \quad \text{and} \quad \tan \Theta = \frac{\sin \Theta_0}{\sin \Theta_1}$$

For practical applications, one can use $M = N$, because the stability factor is not very sensitive to the M/N ratio. Notice that SF_0 from Equation (7.8a) is used in all cases when $\lambda \geq 0$; the stability factor SF_{01} from Equation (7.8b) is applicable only when both

$$\lambda < 0 \quad \text{and} \quad l_3 F_D \cos \delta + l_1 F_S \sqrt{1 - a_\Theta^2} \cos \beta < 0$$

A particle is stable when $SF_0 > 1$ and is unstable when $SF_0 < 1$. Incipient motion corresponds to $SF_0 = 1$. This method reduces to Stevens and Simons' (1971) method for SF_0 when $\Theta_0 = 0$. Example 7.1 provides the detailed calculations of particle stability using this method.

Example 7.1   Application of the moment stability of a particle. A round 5-cm particle stands on the bed of a channel. If the downstream channel slope is 0.05 and the sideslope angle is 20°, calculate

(a) the stability factor of the particle under an applied shear, $\tau_0 = 1$ lb/ft², when the streamlines are deflected downward at a 20° angle; and

(b) the direction of the path line if the particle in (a) enters motion.

(a) Stability factor

1. The particle size is 5 cm $= 0.164$ ft;
2. the angle of repose is approximately $\phi = 37°$ (from Fig. 7.2);
3. the sideslope angle is $\Theta_1 = 20°$,
4. the downstream slope angle $\Theta_0 = \tan^{-1} 0.05 = 2.86°$;
5. the angle $\Theta = \tan^{-1}(\sin \Theta_0 / \sin \Theta_1) = 8.3°$;
6. the factor $a_\Theta = \sqrt{\cos^2 \Theta_1 - \sin^2 \Theta_0} = 0.938$;
7. the deviation angle $\lambda = 20°$, angle positive downward;
8. from Equation (7.11a),

$$\eta_0 = \frac{21 \times 1\,\text{lb}}{\text{ft}^2}\,\frac{\text{ft}^3}{1.65 \times 62.4\,\text{lb} \times 0.164\,\text{ft}} = 1.24$$

9. from Equation (7.13), assuming $M = N$,

$$\beta = \tan^{-1}\left\{\cos(20° + 8.3°)\bigg/\left[2\frac{\sqrt{1-(0.94)^2}}{1.24\tan 37°} + \sin(20° + 8.3°)\right]\right\}$$

$$= 35.9°$$

10. from Equation (7.10),

$$\eta_1 = 1.24\left\{\frac{1 + \sin(20° + 35.9° + 8.3°)}{2}\right\} = 1.18$$

11. from Equation (7.8a), because $\lambda \geq 0$,

$$SF_0 = \frac{0.938\tan 37°}{1.18\tan 37° + \sqrt{1-(0.938)^2}\cos 35.9°} = 0.604$$

The particle is unstable because $SF_0 < 1$.

(b) Direction of the path line

The particle will move along a path line at an angle $\Theta + \beta = 44.2°$ from the vertical in the downstream direction. The angle β is measured on the embankment plane. Its projection in a horizontal plane β_h is given by

$$\beta_h = \tan^{-1}\left[\frac{\tan\beta}{\cos\Theta_1}\right] = \tan^{-1}\left(\frac{\tan 35.9}{\cos 20°}\right) = 37.6°$$

7.4   Simplified stability analysis

An approximate formulation of particle stability is presented as the ratio of critical shear stress on an embankment slope $\tau_{\Theta c}$ compared with the critical shear stress on a flat surface τ_c. If the ratio of forces is used, the angle of repose ϕ can be expressed as a function of the resultant destabilizing force F_{R1} and the stabilizing force $F_S a_\Theta$. With reference to Figure 7.10, the magnitude of the destabilizing force is equal to the square root of the sum of the three squared orthogonal force components:

$$\tan\phi = \frac{F_{R1}}{F_S a_\Theta}$$

$$= \frac{[(F_D \cos\lambda + F_S \sin\Theta_0)^2 + (F_D \sin\lambda + F_S \sin\Theta_1)^2 + F_L^2]^{1/2}}{F_S a_\Theta} \tag{7.14}$$

where the drag force at an angle Θ can be expressed as $F_D = c_2\tau_{\Theta c}d_s^2$.

For incipient motion ($\tau_{\Theta c} = \tau_c$) on a horizontal surface ($\Theta_1 = \Theta_0 = \lambda = 0$), consider the drag force $F_D = c_2\tau_c d_s^2$ and a constant value of the lift-drag ratio $\Pi_{ld} = F_L/F_D$. Equation (7.14) reduces to

$$\tan\phi = \frac{(c_2^2\tau_c^2 d_s^4 + F_L^2)^{1/2}}{F_S} = \frac{c_2\tau_c d_s^2}{F_S}\sqrt{1+\Pi_{ld}^2} \tag{7.15}$$

After combining Equations (7.14) and (7.15) and solving for $\tau_{\Theta c}/\tau_c$, it follows that

$$\frac{\tau_{\Theta c}}{\tau_c} = -\frac{\sin\Theta_1\sin\lambda + \sin\Theta_0\cos\lambda}{\sqrt{1+\Pi_{ld}^2}\tan\phi}$$

$$+ \sqrt{\frac{(\sin\Theta_1\sin\lambda + \sin\Theta_0\cos\lambda)^2}{(1+\Pi_{ld}^2)\tan^2\phi} + 1 - \left(\frac{\sin^2\Theta_0 + \sin^2\Theta_1}{\sin^2\phi}\right)} \tag{7.16} \blacklozenge$$

This equation expresses the ratio of the critical shear stress on the embankment, $\tau_{\Theta c}$, to the critical shear stress on a horizontal surface, τ_c, as a function of (1) the embankment slopes Θ_0 and Θ_1, (2) the shear stress direction angle λ, (3) the angle of repose ϕ, and (4) the lift–drag ratio Π_{ld}. After expanding the square-root term of Equation (7.16) into power series, it follows that

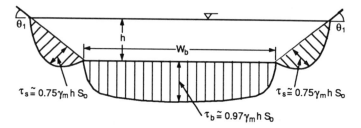

Figure 7.11. Applied shear stress distribution in trapezoidal channels

$$\frac{\tau_{\Theta c}}{\tau_c} = \sqrt{1 - \frac{\sin^2 \Theta_1}{\sin^2 \phi} - \frac{\sin^2 \Theta_0}{\sin^2 \phi}\left(1 - x + \frac{x^2}{2} + \cdots\right)} \tag{7.17a}$$

where

$$x = \frac{\cos \phi (\sin \Theta_1 \sin \lambda + \sin \Theta_0 \cos \lambda)}{\sqrt{1 + \Pi_{ld}^2}\sqrt{\sin^2 \phi - \sin^2 \Theta_1 - \sin^2 \Theta_0}} \tag{7.17b}$$

Consider the particular case where $\Theta_0 = 0$. As the lift forces become negligible ($\Pi_{ld} = 0$), Equation (7.16) reduces to Brooks' relationship. As the lift–drag ratio goes to infinity ($\Pi_{ld} \to \infty$) or when the streamlines are horizontal ($\lambda = 0$), Equation (7.17) with $x = 0$ reduces to Lane's relationship:

$$\frac{\tau_{\Theta c}}{\tau_c} = \cos \Theta_1 \sqrt{1 - \left(\frac{\tan^2 \Theta_1}{\tan^2 \phi}\right)} = \sqrt{1 - \left(\frac{\sin^2 \Theta_1}{\sin^2 \phi}\right)} \tag{7.18}$$

This very simple relationship has become useful for defining the cross-sectional geometry of straight channels. Consider steady uniform flow in a straight trapezoidal channel for a base width–depth ratio $W_b/h > 4$, as shown in Figure 7.11. The bed shear stress $\tau_b \cong 0.97\gamma_m hS_0$ and the bank shear stress $\tau_s \cong 0.75\gamma_m hS_0$ impose two conditions for channel stability: (1) The bed particles are stable when $\tau_b < \tau_c$; and (2) the sideslope particles are stable when $\tau_s < \tau_{\Theta c}$ calculated from Equation (7.18). Table 7.4 suggests sideslope angles Θ_1 for a variety of channel bank material. A detailed stable channel design procedure is presented in Example 7.2.

Example 7.2 Design of a stable trapezoidal channel. Design a stable straight trapezoidal channel made of very coarse rounded gravel, $d_{50} = 1.5$ in. and $d_{90} = 2$ in. A total clear-water discharge $Q = 1,000$ ft³/s is conveyed on a bed slope $S_0 = 0.0015$.

Step 1. The angle of repose $\phi = 37°$ is found from Figure 7.2 for rounded material. The flow depth is set such that the bed particles are at incipient

Table 7.4. *Suggested embankment side slope angles*

Bank material	Θ_1 (deg)
Rock	78.7
Smooth or weathered rock, shell	45–63
Soil (clay, silt, and sand mixtures)	34
Sandy soil	34
Silt and loam (loose sandy earth)	26
Fine sand	18
Other very fine material	18
Compacted clay	34

motion, assuming $R_h = h$. The bed shear stress τ_b equals the critical shear stress τ_c at the following flow depth h:

$$\tau_b = 0.97\gamma h S_0 = 0.06(\gamma_s - \gamma)d_{50}\tan\phi$$

$$h = \frac{0.06(G-1)}{0.97}\frac{d_{50}}{S_0}\tan\phi = \frac{0.06 \times 1.65 \times 1.5 \text{ ft} \times \tan 37°}{0.97 \times 0.0015 \times 12} = 6.3 \text{ ft}$$

Step 2. The sideslope angle Θ_1 is calculated to correspond to beginning of motion, from Equation (7.18). The applied bank shear stress τ_s equals the critical shear stress $\tau_{\theta c}$ at a sideslope angle Θ_1 given by

$$\tau_s = 0.75\gamma h S_0 = 0.06\tan\phi(\gamma_s - \gamma)d_{50}\sqrt{1 - \frac{\sin^2\Theta_1}{\sin^2\phi}}$$

$$\sqrt{1 - \frac{\sin^2\Theta_1}{\sin^2 37°}} = \frac{0.75 \times 6.3 \text{ ft} \times 0.0015 \times 12}{0.06 \times 1.65 \times 1.5 \text{ ft} \times \tan 37°} = 0.76$$

$$\sin^2\Theta_1 = \sin^2 37°(1 - 0.76^2)$$

$$\Theta_1 = 23°$$

Step 3. The cross-sectional area of a trapezoidal channel is given by $A = W_b h + h^2\cot\Theta_1 = W_b h + 93.5 \text{ ft}^2$, given the base width W_b. The wetted perimeter $P = W_b + 2h\csc\Theta_1 = W_b + 32.2 \text{ ft}$, and the hydraulic radius $R_h = A/P$. With the Darcy–Weisbach friction factor from Equation (6.14), the velocity is given by

$$V = \frac{Q}{A} = \sqrt{\frac{8g}{f}}R_h^{1/2}S_0^{1/2} = \sqrt{gR_h S_0}\left[5.75\log\left(\frac{12.2h}{3d_{90}}\right)\right]$$

$$= 2.76\sqrt{R_h}$$

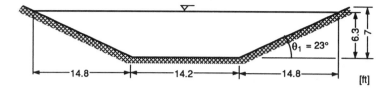

Figure E7.2.1. Stable trapezoidal channel

As a first approximation, $A = (Q/V) \cong W_b h$ with $R_h \cong h = 6.3$ ft gives $W_b \approx 22.9$ ft. A second iteration from $W_b \cong 22.9$ ft gives $A = W_b h + 93.5 = 237$ ft^2, $P = W_b + 32.2 = 55$ ft, $R_h = A/P = 4.3$ ft, $V = 2.76\sqrt{R_h} = 5.72$ ft/s, and $W_b = [(Q/V) - 93.5]\, 1/h = 12.9$ ft. A third iteration from $W_b = 12.9$ ft gives $A = 175$ ft^2, $P = 45.1$ ft, $R_h = 3.9$ ft, $V = 5.43$ ft/s, $W_b = 14.4$ ft. A fourth iteration from $W_b = 14.4$ ft gives $A = 184$ ft^2, $P = 46.6$ ft, $R_h = 3.94$ ft, $V = 5.48$ ft/s, $W_b = 14.1$ ft. The base width should be about $W_b = 14.2$ ft for a stable channel. The cross-sectional geometry is plotted in Figure E7.2.1.

Lane (1953) defined the at-a-station geometry of a straight canal cross section for which all particles are simultaneously at incipient motion. Assuming that at any point the rectilinear shear stress in the downstream direction equals the shear stress of the column of the water–sediment mixture above it, the solution to Equation (7.18) leads to the cosine cross section discussed in Example 7.3.

Example 7.3 Analysis of ideal cross-sectional geometry. Consider steady uniform flow in a straight channel, $\lambda = 0$, of uniform grain size with a negligible downstream bed slope $\Theta_0 \rightarrow 0$. Equation (7.18) indicates that the critical shear stress for beginning of motion on the channel sideslopes at an angle Θ_1 is always less than on a flat surface. Considering that the applied shear stress on the sideslopes of a channel is also less than the applied shear stress at the lowest point of a cross section, it is possible to define the cross-sectional geometry of a channel for which all the particles along its wetted perimeter simultaneously reach incipient motion.

The applied boundary shear stress at any point along the wetted perimeter being approximately equal to $\tau_0 = \gamma h S_0$, the profile is calculated from Equation (7.18) given that $\tan \Theta_1 = -dh/dy$. The solution derived by Lane (1953) is a cosinusoidal cross section, as shown in Figure E7.3.1.

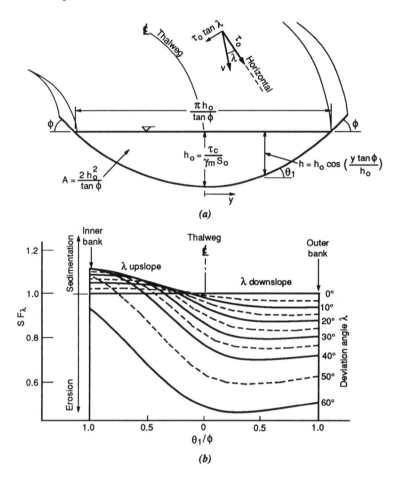

Figure E7.3.1. (a) Ideal cross-sectional geometry and (b) relative particle stability

The maximum flow depth is $h_0 = \tau_c/\gamma_m S_0$, the channel width is $W = \pi h_0/\tan\phi$, and the cross-sectional area is $2h_0^2 \cot\phi$.

Granted that all the particles on the wetted perimeter are at incipient motion, or $SF_0 = 1$, under straight downstream flow conditions, it is instructive to examine the relative stability of the same particles when the streamlines are deflected at an angle λ from the downstream direction due to secondary circulation in a bend. At a given applied boundary shear stress τ_0, the particle stability under streamlines deflected at an angle λ is compared with the particle stability for horizontal streamlines $\lambda = \Theta_2 = 0$. The analysis is pursued in two steps.

First, given constant values of d_s, γ_s, γ, $\Theta_2 = 0$, ϕ, $\lambda = 0$ and assuming $M = N$, a trial value of τ_0 is assumed for parametric values of Θ_1. The plane bed stability number η_0 is calculated from Equation (7.11), the angle β from Equation (7.13), the sideslope stability number η_1 from Equation (7.10), and the stability factor SF_0 from Equation (7.8a). This procedure is repeated for different values of τ_0 until SF_0 equals unity. In parametric form, a reference set of τ_0 values corresponding to $SF_0 = 1$ is obtained for parametric values of Θ_1/ϕ.

Second, deviation angles $\lambda \neq 0$ are considered. A transverse shear stress component $\tau_0 \tan \lambda$ is applied in the direction normal to the downstream shear stress component from the first step. The new stability factor SF_λ with $\lambda \neq 0$ is calculated successively from Equations (7.11a), (7.13), (7.10), and (7.8a); the results are shown in Figure E7.3.1b. Obviously, as $\lambda = 0$, $SF_\lambda = 1$ for all values of Θ_1/ϕ.

It is concluded that when the deviation angle λ is small, the particles near the outer bank become unstable ($SF_\lambda < 1$) while those near the inner bank become stable ($SF_\lambda > 1$). This induces erosion near the outer bank of the cross section and sedimentation near the inner bank of the cross section. As the angle λ increases, the proportion of the cross section under erosion increases, and the entire cross section is eroded as $\lambda > 50°$. Equilibrium between zones of erosion and sedimentation in bends with secondary flow ($\lambda \neq 0$) is therefore possible at low deviation angles λ ($\lambda < \sim 15°$).

The downstream hydraulic geometry of noncohesive alluvial channels in terms of width W, average flow depth $\tilde{h}$, average flow velocity $\tilde{V}$, and friction slope S_f can then be analyzed by combining the following four fundamental equations:

$$Q = W\tilde{h}\tilde{V} \tag{E7.3.1}$$

$$\tilde{V} = b_v \sqrt{8g}\left(\frac{\tilde{h}}{d_s}\right)^{1/6} \tilde{h}^{1/2} S_f^{1/2} \tag{E7.3.2}$$

$$\tau_{\Theta*} = \frac{k_r \tilde{h} S_f}{(G-1)d_s} \tag{E7.3.3}$$

$$\tan \lambda = b_r \left(\frac{\tilde{h}}{d_s}\right)^{1/6} \frac{\tilde{h}}{R_c} \tag{E7.3.4}$$

where Q is the total discharge, b_v is an empirical resistance coefficient, g is the gravitational acceleration, d_s is the median grain size, ρ and ρ_s are, respectively, the mass densities of water and sediment, k_r is a coefficient for the hydraulic radius $R_h = k_r\tilde{h}$, λ is the streamline deviation angle, b_r

is a coefficient, and R_c is the channel radius of curvature. These four equations can be combined and solved (Julien, 1988; Wargadalam, 1993) to define the downstream hydraulic geometry:

$$W = 0.76 Q^{0.5} d_s^{-0.28} \tau_{\Theta*}^{-0.32} \qquad (E7.3.5)$$

$$\tilde{h} = 0.37 Q^{0.34} d_s^{-0.004} \tau_{\Theta*}^{0.07} \qquad (E7.3.6)$$

$$\tilde{V} = 6.0 Q^{0.12} d_s^{0.30} \tau_{\Theta*}^{0.28} \qquad (E7.3.7)$$

$$S_f = 6.7 Q^{-0.37} d_s^{0.95} \tau_{\Theta*}^{1.07} \qquad (E7.3.8)$$

where the discharge Q is in m^3/s, the median grain size d_{50} is in m, and $\tau_{\Theta*}$ is the Shields parameter.

The hydraulic geometry of noncohesive alluvial channels therefore varies primarily with discharge Q and grain size d_s. It is noticeable that the width and depth increase primarily with discharge. The slope depends largely on grain size and varies inversely with discharge. Interestingly, the velocity also increases with grain size.

---

*Problem 7.1

What is the sediment size corresponding to beginning of motion when the shear velocity $u_* = 0.1$ m/s?

Answer: medium gravel, $d_s \cong 1$ cm.

*Problem 7.2

Given the stream slope $S_0 = 1 \times 10^{-3}$, at what flow depth would coarse gravel enter motion?

**Problem 7.3

Calculate the stability factor of 8-in. riprap on an embankment inclined at a $1V:2H$ sideslope if the shear stress $\tau_0 = 1$ lb/ft^2.

Answer: SF$_0 = 1.3$; thus, stable.

**Problem 7.4

An angular 10-mm sediment particle is submerged on an embankment inclined at $\Theta_1 = 20°$ and $\Theta_2 = 0°$. Calculate the critical shear stress from the moment stability method when the streamlines near the particle are (a) $\lambda = 15°$ (deflected downward); (b) $\lambda = 0°$ (horizontal flow); and (c) $\lambda = -15°$ (deflected upward).

**Problem 7.5

Compare the values of critical shear stress $\tau_{\theta c}$ from Problem 7.4 with those calculated with Equation (7.16) and with Lane's method [Eq. (7.18)], given $\Theta_0 = 0$ and $\Pi_{ld} = 0$.

Answer: The values are given in the following tabulation:

Angle λ (deg)	Moment stability (N/m²)	Simplified stability (N/m²)	Lane's method (N/m²)
−15 up	6.3	7.0	6.3
0	5.5	6.3	6.3
15 down	4.8	5.7	6.3

*Problem 7.6

Design a stable channel conveying 14 m³/s in coarse gravel, $d_{50} = 10$ mm and $d_{90} = 20$ mm, at a slope $S_0 = 0.0006$.

8

Bedforms

As soon as sediment particles enter motion, the random patterns of erosion and sedimentation generate very small perturbations of the bed surface elevation. In many instances, these perturbations grow until various surface configurations called bedforms cover the entire bed surface (Sections 8.1 and 8.2). Resistance to flow (Section 8.3), which depends largely on bedform configuration, directly affects water surface elevation in alluvial channels. Changes in bedform resistance induce shifting of the stage-discharge relationship (Section 8.4) and create problems in the determination of river discharges from water level measurements.

8.1   Mechanics of bedforms

A fundamental understanding of the mechanics of bedforms can be deduced from an analysis of the equations of motion in the downstream x and upward z directions [Eqs. (3.19a,c)] for steady flow conditions:

$$\frac{\partial}{\partial x}\left(\frac{p}{\rho_m} + \frac{v^2}{2}\right) = g_x + (v_y \otimes_z - v_z \otimes_y) + \frac{1}{\rho_m}\left(\frac{\partial \tau_{xx}}{\partial x} + \frac{\partial \tau_{yx}}{\partial y} + \frac{\partial \tau_{zx}}{\partial z}\right) \tag{8.1}$$

$$\frac{\partial}{\partial z}\left(\frac{p}{\rho_m} + \frac{v^2}{2}\right) = g_z + (v_x \otimes_y - v_y \otimes_x) + \frac{1}{\rho_m}\left(\frac{\partial \tau_{xz}}{\partial x} + \frac{\partial \tau_{yz}}{\partial y} + \frac{\partial \tau_{zz}}{\partial z}\right) \tag{8.2}$$

Assuming two-dimensional flow in a wide open channel, (1) the shear stress components τ_{xx}, τ_{yx}, τ_{yz}, and τ_{zz} can be neglected as a first approximation, and (2) the velocity components v_y and v_z are assumed to remain negligible such that $v_x = v$.

With the only rotation component $\otimes_y = \partial v_x / \partial z$ obtained from Equation (3.4), and the specific energy function $E = p/\gamma_m + v^2/2g$ from Equation (3.23a), the equations of motion [Eqs. (8.1) and (8.2)] reduce, respectively, to

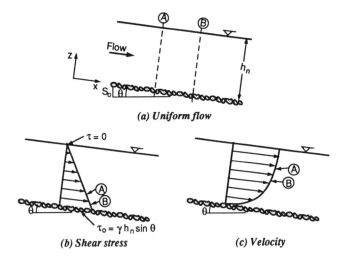

(a) Uniform flow

(b) Shear stress   (c) Velocity

Figure 8.1. Steady uniform flow in alluvial channels

$$g \frac{\partial}{\partial x} \left[\frac{p}{\rho_m g} + \frac{v^2}{2g} \right] = g_x + \frac{1}{\rho_m} \frac{\partial \tau_{zx}}{\partial z} \tag{8.3}$$

$$\frac{\partial}{\partial z} \left[\frac{p}{\rho_m} + \frac{v^2}{2} \right] = g_z + v_x \frac{\partial v_x}{\partial z} + \frac{1}{\rho_m} \frac{\partial \tau_{xz}}{\partial x} \tag{8.4}$$

The pressure distribution remains hydrostatic when $\partial \tau_{xz}/\partial x = 0$, and Equation (8.4) can be easily depth-integrated over h_n with $g_z = -g \cos \theta$ to give

$$p = \rho_m g(h_n - z) \cos \theta \tag{8.5}$$

For steady uniform flow (Fig. 8.1a), the velocity v and the pressure p remain constant along x and the left-hand side of Equation (8.3) reduces to zero. The shear stress distribution τ_{zx} is then obtained after integrating the right-hand side of Equation (8.3) over the normal depth h_n with $g_x = g \sin \theta$; thus,

$$\tau_{zx} = \rho_m g(h_n - z) \sin \theta \tag{8.6}$$

Note that, for steady uniform flow, the shear stress increases linearly from $\tau_{zx} = 0$ at the free surface to the boundary shear stress $\tau_0 = \tau_{zx} = \rho_m g h_n \sin \theta = \gamma_m h_n S_0$ at the bed.

In the case of nonuniform flow with hydrostatic pressure distribution, the governing equation describing the internal shear stress distribution [Eq. (8.3)] can be rewritten as

$$-\frac{1}{\rho_m}\frac{\partial \tau_{zx}}{\partial z} = \underbrace{g\sin\theta}_{\text{uniform flow}} - \underbrace{g\frac{\partial}{\partial x}\left(\frac{p}{\rho_m g}+\frac{v^2}{2g}\right)}_{\text{nonuniform flow perturbation}} \quad \text{throughout the flow} \quad (8.7a)$$

The cases of subcritical and supercritical flow are considered separately.

8.1.1   *Subcritical flow*

In subcritical flow, consider a small bed perturbation of amplitude Δz, as sketched in Figure 8.2a. Two points, C and D, are identified where the flow depth is h_n on each side of the perturbation for comparison with shear stresses for steady uniform flow.

Considering the entire flow depth, it is shown from the specific energy diagram in Figure 8.2b that a small perturbation Δz causes a decrease in specific energy when approaching the perturbation; thus, $g\,\partial E/\partial x < 0$ at point C. It follows that the gradient of shear stress $\partial \tau_{zx}/\partial z$ on the upstream side becomes larger (positive nonuniform flow perturbation) than for the corresponding steady uniform flow condition. Conversely,

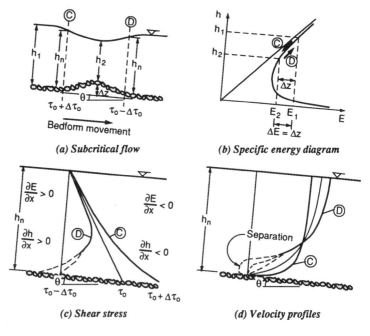

(a) Subcritical flow   *(b) Specific energy diagram*

(c) Shear stress   *(d) Velocity profiles*

Figure 8.2. Steady nonuniform subcritical flow in alluvial channels

on the downstream side of the perturbation, the corresponding increase in specific energy ($g\,\partial E/\partial x > 0$) causes a reduction in shear stress (reduced $\partial \tau_{zx}/\partial z$).

Near the bed, however, the velocity term within parentheses in Equation (8.7a) can be neglected and the shear stress gradient near the bed reduces to

$$-\frac{1}{\rho_m}\frac{\partial \tau_{zx}}{\partial z} = g\sin\theta - g\frac{\partial h}{\partial x} \quad \text{near the bed} \tag{8.7b}$$

On the downstream side of the perturbation, the gradient of shear near the bed becomes positive when $\partial h/\partial x$ exceeds $\sin\theta$, as shown in Figure 8.2c for curve D. Integration over the entire flow depth may result in negative values of bed shear stress τ_0 when $\partial h/\partial x$ becomes large. Separation occurs when $\tau_0 < 0$ on the downstream side of the perturbation, as in Figure 8.2d.

As a consequence of increased shear stress on the upstream face of the perturbation in subcritical flows, increased sediment transport causes erosion in converging flow. On the downstream side of the perturbation, the reduced bed shear stress and sediment transport capacity induces sedimentation on the lee side of the perturbation. This mechanism causes the amplification of the perturbation until large bedforms fully develop.

8.1.2 Supercritical flow

In the case of supercritical flow, a similar analysis is depicted in Figure 8.3. The small perturbation causes an increase in flow depth and a reduction in specific energy near F. Upstream of the perturbation, this results in a steeper shear stress gradient in the upper part of the velocity profile and a reverse gradient near the bed (Fig. 8.3c), which may induce separation of the velocity profile near the bed (Fig. 8.3d). Conversely, the downstream face of the perturbation at G shows an increase in boundary shear stress, which increases sediment transport capacity on the downstream side of the perturbation. Although the sediment particles are transported in the downstream direction, bedforms in supercritical flow migrate upstream as the sediment deposition on the upstream face combines with erosion on the downstream face of the perturbation.

The celerity c of small-amplitude gravity waves represents the ratio of the wavelength λ to the wave period T_w. The celerity of surface waves is given as

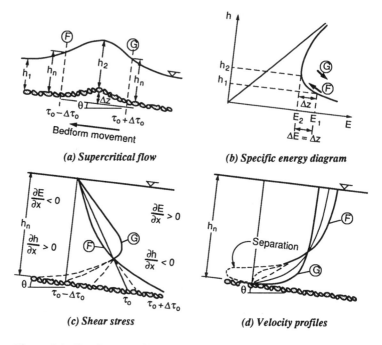

(a) Supercritical flow

(b) Specific energy diagram

(c) Shear stress

(d) Velocity profiles

Figure 8.3. Steady nonuniform supercritical flow in alluvial channels

$$c = \frac{\lambda}{T_w} = \sqrt{\frac{g\lambda}{2\pi} \tanh\left(\frac{2\pi h}{\lambda}\right)} \qquad (8.8)$$

Gravity waves are classified according to the water depth h in which they travel. Deep-water waves, $\lambda < 2h$, travel at a celerity $c = \sqrt{g\lambda/2\pi}$. Shallow-water waves, $\lambda > 25h$, travel at a constant celerity $c = \sqrt{gh}$. In the transitional region, $2h < \lambda < 25h$, the celerity is given by Equation (8.8). In the case where $\lambda = 2\pi h$, $\tanh 1 = 0.7616$, and $c = 0.873\sqrt{gh}$. The mean flow velocity V equals the wave celerity c when $\mathrm{Fr} = 1$ for shallow-water waves ($\lambda > 25h$); $\mathrm{Fr} = 0.873$ when $\lambda = 2\pi h$; and $\mathrm{Fr} \to 0$ as $\lambda \to 0$.

8.2   Bedform classification and geometry

Various bedform configurations and geometry define the boundary roughness and resistance to flow in alluvial channels. The primary variables that affect bedform configuration and geometry are the slope of the energy grade line, flow depth, bed particle size, and particle fall velocity.

Flat bed, or plane bed, refers to a bed surface without bedforms. With reference to the bedform configurations in Figure 8.4, ripples are small

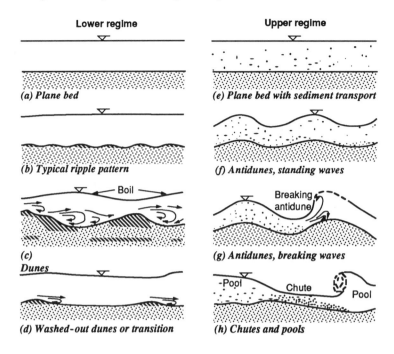

Figure 8.4. Types of bedforms in alluvial channels
(after Simons and Richardson, 1966)

bedforms with wave heights less than ~0.1 ft. Ripple shapes vary from nearly triangular to almost sinusoidal. Dunes are larger than ripples and are out of phase with the water surface waves. From longitudinal profiles, dunes are often triangular with fairly gentle upstream slopes and downstream slopes approaching the angle of repose of the bed material. The large eddies on the lee side of dunes may cause surface boils. Transition from lower regime to upper regime consists of low-amplitude ripples, washed-out dunes, and plane bed with sediment transport. Antidunes, or standing waves, are in phase with free-surface waves. They grow with increasing Froude number until they become unstable and break. Chutes and pools occur at relatively large slopes and consist of elongated chutes with supercritical flow connected by pools where the flow is generally subcritical.

Bedforms are classified into lower and upper flow regimes based upon their shape, resistance to flow, and mode of sediment transport (Simons and Richardson, 1963, 1966). A transition zone exists between the two flow regimes, where bedforms range from washed-out dunes to plane bed

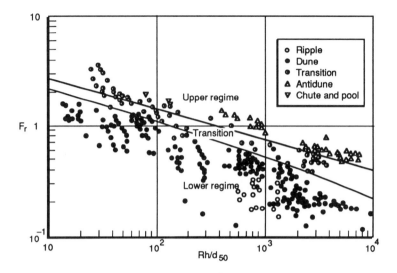

Figure 8.5. Lower- and upper-regime bedform classification
(after Athaullah, 1968)

or standing waves. The Froude numbers in the transition zone have been
found by Athaullah (1968) to decrease with relative submergence, R_h/d_{50},
as shown in Figure 8.5.

In the lower flow regime, the flow is generally subcritical (Fr < 1) and
the water surface undulations are out of phase with the bed waves (Fig.
8.4). Surface roughness characteristics are summarized in Table 8.1. Resistance to flow is large because flow separation occurs on the downstream
side of the waves. This generates large-scale turbulence, dissipating considerable energy. Sediment transport is relatively low because bed sediment particles move primarily in contact with the bed. Dunes measuring
30–60 ft in height and several hundred feet in length have been observed
in deep rivers.

In the upper flow regime, resistance to flow is low because grain roughness predominates. However, the energy dissipated by standing waves
and the formation of breaking antidunes increases flow resistance. Standing waves and antidunes are common in supercritical flow (Fr > 1). Standing waves are in-phase sinusoidal sand and water waves (Fig. 8.5) that
build up in amplitude from a plane bed and plane water surface, then
slowly fade away. Resistance to flow for breaking antidunes is slightly
larger than for plane bed, as they cover only a small portion of channel
reach at a given time. Bed material transport in the upper flow regime is
high because, except when antidunes are breaking, the contact sediment

Table 8.1. *Typical bedform characteristics*

Bedform	Manning coefficient n	Concentration (mg/l)	Dominant type of roughness	Bedform surface profiles
Lower flow regime				
Plane bed	0.014	0	Grain	—
Ripples	0.018–0.028	10–200	Form	—
Dunes	0.020–0.040	200–3,000	Form	Out of phase
Washed-out dunes	0.014–0.025	1,000–4,000	Variable	Out of phase
Upper flow regime				
Plane bed	0.010–0.013	2,000–4,000	Grain	—
Antidunes	0.010–0.020	2,000–5,000	Grain	In phase
Chutes and pools	0.018–0.035	5,000–50,000	Variable	In phase

discharge is almost continuous and the suspended sediment concentration is large.

The prediction of bedform configurations has been the subject of numerous laboratory and field investigations. Liu (1957) used the ratio u_*/ω of the shear velocity u_* to the particle fall velocity ω as a function of the grain shear Reynolds number $Re_* = u_* d_s/\nu_m$. His analysis in Figure 8.6 suggests that ripples can form only on hydraulically smooth beds (including

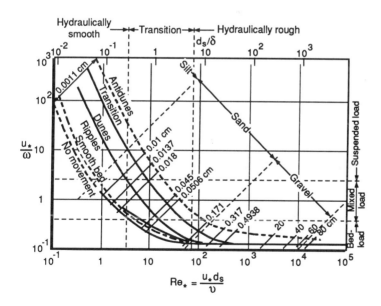

Figure 8.6. Bedform classification (after Liu, 1957)

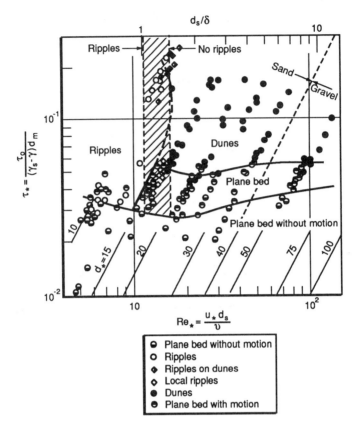

Figure 8.7. Bedform classification (after Chabert and Chauvin, 1963)

transition). Chabert and Chauvin (1963) proposed a bedform predictor based on the Shields diagram, shown in Figure 8.7. Ripples form when $d_* = d_s[(G-1)g/\nu_m^2]^{1/3} < 15$, which corresponds to $d_s \leq 0.6$ mm.

Simons and Richardson (1963, 1966) proposed a bedform predictor encompassing both lower and upper regimes when plotting the stream power $\gamma q S_f$ as a function of particle diameter (Fig. 8.8). Their bedform predictor based on extensive laboratory experiments is quite reliable for shallow streams but deviates from observed bedforms in deep streams. Bogardi (1974) defined a particle stability factor gd_s/u_*^2 that bears resemblance to the Shields number $\tau_* = 0.6u_*^2/gd_s$. The predictor shown in Figure 8.9 is also similar to Liu's predictor (Fig. 8.6).

Van Rijn (1984b) proposed a bedform classification based on the dimensionless particle diameter d_* and the transport-stage parameter T,

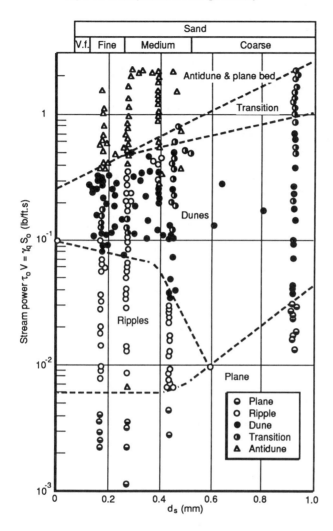

Figure 8.8. Bedform classification (after Simons and Richardson, 1963, 1966)

defined, respectively, as

$$d_* = d_{50}\left[\frac{(G-1)g}{\nu_\mathrm{m}^2}\right]^{1/3} \tag{8.9a}$$

$$T = \frac{\tau_*' - \tau_{*\mathrm{c}}}{\tau_{*\mathrm{c}}} = \frac{(u_*')^2 - u_{*\mathrm{c}}^2}{u_{*\mathrm{c}}^2} = \frac{\rho_\mathrm{m} V^2}{\tau_\mathrm{c}[5.75\log(4R_\mathrm{b}/d_{90})]^2} - 1 \tag{8.9b}$$

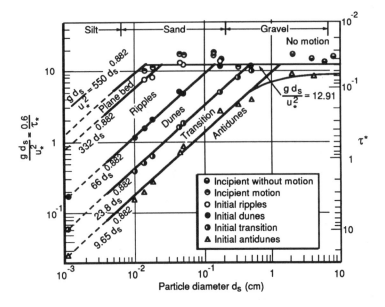

Figure 8.9. Bedform classification (after Bogardi, 1974)

in which d_{50} is the median bed particle diameter (50% passing by weight), G the particle specific gravity, ν_m the fluid mixture kinematic viscosity, V the depth-averaged flow velocity, g the gravitational acceleration, R_b the hydraulic radius related to the bed obtained from the Vanoni–Brooks method (Example 6.2), d_{90} the 90% passing bed particle diameter, and τ_c the critical shear stress obtained from the Shields diagram. The parameters τ_*' and u_*' are discussed in Section 8.3.

Van Rijn suggested that the ripples form when both $d_* < 10$ and $T < 3$, as shown in Figure 8.10. Dunes are present elsewhere when $T < 15$, dunes wash out when $15 < T < 25$, and upper flow regime starts when $T > 25$.

In the lower regime, the geometry of bedforms refers to representative dune height Δ and wavelength Λ as a function of the average flow depth h, median bed particle diameter d_{50}, and other flow parameters such as the transport-stage parameter T, and the grain shear Reynolds number $\mathrm{Re}_* = u_* d_s / \nu_m$. The bedform height and steepness predictors proposed by van Rijn (1984b) are

$$\frac{\Delta}{h} = 0.11\left(\frac{d_{50}}{h}\right)^{0.3}(1 - e^{-0.5T})(25 - T) \qquad (8.10a)$$

and

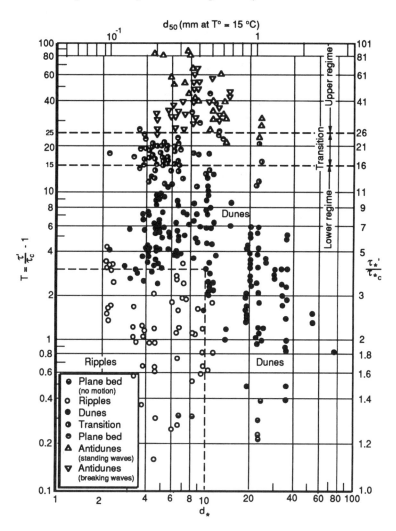

Figure 8.10. Bedform classification (after van Rijn, 1984b)

$$\frac{\Delta}{\Lambda} = 0.015\left(\frac{d_{50}}{h}\right)^{0.3}(1 - e^{-0.5T})(25 - T) \tag{8.10b}$$

The bedform length obtained from dividing these two equations, $\Lambda = 7.3h$, is close to the theoretical value, $\Lambda = 2\pi h$, derived by Yalin (1964). The agreement with laboratory data is quite good, as shown in Figure 8.11a,b, but both curves tend to underestimate the bedform height and

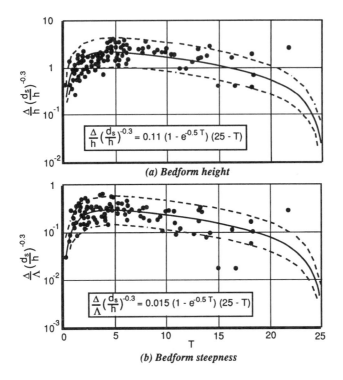

(a) Bedform height

(b) Bedform steepness

Figure 8.11. Bedform height and steepness (after van Rijn, 1984b)

steepness of field data (Julien, 1992). Lower-regime bedforms are observed in the Mississippi River at values of T well beyond 25.

8.3   Resistance to flow with bedforms

The analysis of total resistance is somewhat analogous to the analysis of viscous flow around a spherical particle presented in Section 5.3. The total resistance is separated into (1) grain resistance accounting for forces acting on individual particles and (2) form resistance due to bedform configurations. The total bed shear stress is the sum of two components,

$$\tau_b = \tau_b' + \tau_b'' \tag{8.11a}$$

where τ_b is the total bed shear stress, τ_b' the grain shear stress, and τ_b'' the form shear stress.

The corresponding identities using the grain shear velocity u_*', grain hydraulic radius R_h', grain friction slope S_f', grain Darcy–Weisbach friction factor f', and the corresponding values u_*'', R_h'', S_f'', and f'' for the form resistance are formulated as

$$u_*^2 = u_*'^2 + u_*''^2 \tag{8.11b}$$

$$R_h = R_h' + R_h'' \tag{8.11c}$$

$$S_f = S_f' + S_f'' \tag{8.11d}$$

$$f = f' + f'' \tag{8.11e}$$

When written in terms of Shields parameters, Equation (8.11a) gives $\tau_* = \tau_*' + \tau_*''$, or

$$\frac{\tau_b}{(\gamma_s - \gamma_m)d_{50}} = \frac{\tau_b'}{(\gamma_s - \gamma_m)d_{50}} + \frac{\tau_b''}{(\gamma_s - \gamma_m)d_{50}} \tag{8.11f}$$

where τ_* is the total Shields parameter, τ_*' the grain Shields parameter, and τ_*'' the form Shields parameter. Notice that, for the Manning and Chézy coefficients, $n \neq n' + n''$ and $C \neq C' + C''$.

The grain shear stress τ_b' describes the shear stress actually applied on bed particles. This component is assumed to determine beginning of motion and sediment transport. Two methods proposed by Engelund (1966) and van Rijn (1984b) can be used to separate grain resistance from form resistance in alluvial channels.

8.3.1 Engelund's method

The relationship between total and grain Shields parameters from laboratory flume data is shown in Figure 8.12. Engelund (1966) estimated S'' from expansion losses in the separation zone downstream from the dune crest. Using the Carnot losses formula, he suggested

$$\Delta H'' = \frac{C_E(V_1 - V_2)^2}{2g} \tag{8.12a}$$

where C_E is the expansion loss coefficient, and V_1 and V_2 are the depth-averaged velocities at the crest and toe of the dune, respectively. Given the dune height Δ, dune length Λ, mean flow depth h, and unit discharge q, one obtains

$$\Delta H'' = \frac{C_E}{2g}\left[\frac{2q}{2h-\Delta} - \frac{2q}{2h+\Delta}\right]^2 \cong C_E \frac{V^2}{2g}\left(\frac{\Delta}{h}\right)^2 \tag{8.12b}$$

or

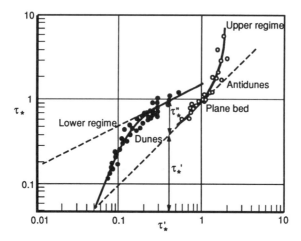

Figure 8.12. Total versus grain resistance
(after Engelund and Hansen, 1967)

$$S'' \cong \frac{C_E}{2} \frac{\Delta^2}{\Lambda h} \mathrm{Fr}^2 \tag{8.13}$$

Accordingly, the bedform energy gradient $S'' = \Delta H''/\Lambda$ depends on the dune steepness Δ/Λ, the relative dune height Δ/h, and the square of the Froude number Fr. Unfortunately, little is known about the expansion loss coefficient C_E. However, except for slight inaccuracies in the location of the transition zone, the results in Figure 8.12 are likely to prevail for sand-bed channels.

8.3.2   Van Rijn's method

Van Rijn (1984b) used the previously defined [Eq. (8.9b)] transport-stage parameter T, which reflects sediment transport with threshold conditions at $T = 0$. The grain shear velocity $u'_* = V\sqrt{g}/C'$, where the grain Chézy coefficient $C' = 5.75\sqrt{g} \log(12R_b/k'_s)$ depends on the average flow velocity V, the bed hydraulic radius R_b, and the surface roughness $k'_s = 3d_{90}$.

In the case of flow over bedforms, van Rijn suggested that the total grain roughness parameter k_s depends not only on grain roughness k'_s, but also on the bedform steepness Δ/Λ, as shown in Figure 8.13. He proposed the following equation:

$$k_s = \underbrace{3d_{90}}_{k'_s} + \underbrace{1.1\Delta(1 - e^{-25\Delta/\Lambda})}_{k''_s} \tag{8.14}$$

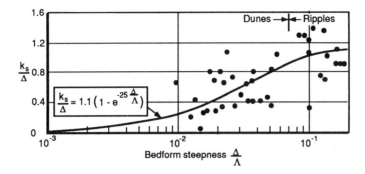

Figure 8.13. Equivalent bedform roughness (after van Rijn, 1984b)

where Δ is the bedform height, Λ the bedform length, and d_{90} the 90% passing bed particle diameter.

The total friction parameter f or total Chézy coefficient C is then calculated from

$$C = \sqrt{\frac{8g}{f}} = 5.75\sqrt{g} \log \frac{12R_b}{k_s} \tag{8.15}$$

where k_s includes the effects of bedforms as given by Equation (8.14).

This relatively recent approach centered around Equation (8.14) offers simplicity but must be tested further with field data. Example 8.1 illustrates how the Engelund and van Rijn methods can be combined to calculate roughness, flow depth, and lower-regime bedform geometry in sandbed channels.

Example 8.1 Application of resistance equations with bedforms. Determine the flow depth using Engelund's method in a 46-m-wide canal in Pakistan conveying 70 m³/s. The slope $S_0 = 11.5$ cm/km and the sediment size is $d_{50} = 0.4$ mm and $d_{90} = 0.65$ mm. Estimate the type and geometry of bedforms using van Rijn's method and verify the flow depth from Equations (8.14) and (8.15).

Flow depth from Engelund's method

Step 1. Assume $h' = 1$ m, large width–depth ratio.

Step 2. Calculate τ'_*:

$$\tau'_* = \frac{\gamma h' S_0}{(\gamma_s - \gamma) d_{50}} = \frac{1 \text{ m} \times 11.5 \times 10^{-5}}{(2.65 - 1) \times 0.0004 \text{ m}} = 0.174$$

Step 3. Determine τ_* from Figure 8.12 with $\tau'_* = 0.175$, $\tau_* \cong 0.55$ in the lower-regime portion of the diagram.

Step 4. Calculate h from τ_*:

$$h = \tau_* \frac{(\gamma_s - \gamma)d_{50}}{\gamma S_0} = 0.55 \frac{(2.65 - 1)0.0004}{11.5 \times 10^{-5}} = 3.15 \text{ m}$$

Step 5. Calculate the mean flow velocity from Equation (6.17) (here van Rijn's equation is arbitrarily selected):

$$V = u'_* \sqrt{\frac{8}{f'}} = \sqrt{gh'S_0} \times 5.75 \log\left(\frac{12h'}{3d_{90}}\right)$$

$$= \sqrt{9.81 \times 1 \times 11.5 \times 10^{-5}} \times 5.75 \log\left(\frac{12 \times 1}{3 \times 0.00065}\right) = 0.732 \text{ m/s}$$

Step 6. Calculate h by continuity:

$$h = \frac{Q}{WV} = \frac{70 \text{ m}^3}{s \times 46 \text{ m} \times 0.73 \text{ m/s}} = 2.08 \text{ m}$$

Step 7. Compare flow depths from Steps 4 and 6, and repeat the procedure assuming different values of h' in Step 1 until both flow depths match. The calculated flow depth is $h' = 0.8$ m, with $\tau'_* = 0.14$, $\tau_* = 0.4$, $h = 2.3$ m, and $V = 0.65$ m/s.

Engelund's method shows that, in the lower regime, most of the energy losses occur because of form roughness. In the case in point, $S = 11.5$ cm/km, $S' = 4$ cm/km, and $S'' = 7.5$ cm/km.

Bedform geometry from van Rijn's method. The type and geometry of bedforms can be obtained from van Rijn's method.

Step 8. Compute d_* from Equation (8.9a), assuming clear water at 20°C:

$$d_* = d_{50}\left[\frac{(G-1)g}{\nu_m^2}\right]^{1/3} = 0.0004\left[\frac{(2.65 - 1) \times 9.81}{1 \times 10^{-12}}\right]^{1/3} = 10.1$$

Step 9. Compute the critical bed shear stress τ_{*c} from Equation (7.4) with $\phi = 30°$:

$$\tau_{*c} = \frac{\rho_m u_{*c}^2}{(\gamma_s - \gamma_m)d_s} = 0.25d_*^{-0.6}\tan\phi = 0.036$$

Step 10. Compute the transport-stage parameter T from Equation (8.9b) and $\tau'_* = 0.14$ calculated in Step 7 of Engelund's method:

$$T = \frac{\tau'_* - \tau_{*c}}{\tau_{*c}} = \frac{0.14 - 0.036}{0.036} = 2.9$$

Step 11. Given the parameter d_* and T, the type of bedform from Figure 8.10 is dunes, although ripples may also be found.

Step 12. Compute bedform height Δ from Equation (8.10a) given the $h = 2.3$ m from Step 7 in Engelund's method:

$$\Delta = 0.11h\left(\frac{d_{50}}{h}\right)^{0.3}(1 - e^{-0.5T})(25 - T)$$

$$\Delta = 0.11 \times 2.3 \text{ m} \times \left(\frac{0.0004}{2.3}\right)^{0.3}(1 - e^{-0.5 \times 2.9})(25 - 2.9)$$

$$\Delta = 0.32 \text{ m}$$

Step 13. Compute bedform length Λ from Equation (8.10b), or $\Lambda = 7.3h$:

$$\Lambda = 7.3h = 7.3 \times 2.3 \text{ m} = 16.7 \text{ m}$$

The last two steps provide a verification of the velocity previously calculated.

Step 14. Compute k_s from Equation (8.14):

$$k_s = 3d_{90} + 1.1\Delta(1 - e^{-25\Delta/\Lambda})$$

$$k_s = 3 \times 0.00065 + 1.1 \times 0.32(1 - e^{(-25 \times 0.32)/16.7}) = 0.136 \text{ m}$$

Step 15. Compute the Chézy coefficient C from Equation (8.15):

$$C = 5.75\sqrt{g} \log\left(\frac{12h}{k_s}\right)$$

$$C = 5.75\sqrt{9.81} \log\left(\frac{12 \times 2.3}{0.136}\right) = 41.6 \frac{\text{m}^{1/2}}{\text{s}}$$

The velocity $V = C\sqrt{hS} = 41.6\sqrt{2.3 \times 11.5 \times 10^{-5}} = 0.67$ m/s is in good agreement with the velocity calculated from Engelund's method (Step 7).

8.4 Field observations

There is no unique relationship between channel slope, flow depth, and flow velocity in alluvial sand-bed channels. Changes in bed configuration are responsible for the nonuniqueness of the stage–discharge relationship. Bed configuration changes affect resistance to flow, flow velocity,

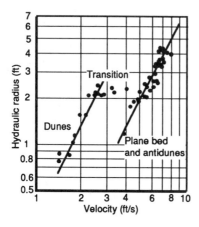

Figure 8.14. Velocity versus hydraulic radius for the Rio Grande (after Nordin, 1964)

and sediment transport. There are many examples of these changes in natural streams. On an approximately 300-mile reach of the Missouri River from Sioux City, Iowa, to Kansas City, Kansas, the bed configurations at a given discharge change from large dunes in the middle of the summer, when water temperature is about 80°F, to washed-out dunes or plane bed in the fall at water temperature of around 45°F. The changes in bedform configuration at constant discharge ($\sim$34,000 ft^3/s) decreased the Manning coefficient n from 0.018 to 0.014 and reduced the average flow depth from $\sim$11 to $\sim$9 ft, with a corresponding increase in average flow velocity from 4.6 to 5.5 ft/s.

The Rio Grande also displays a discontinuous stage–discharge relationship. Nordin (1964) documented the change in hydraulic radius versus mean velocity (Fig. 8.14). During runoff events, the bed varies from dunes at low flow to plane bed and antidunes at high flow, which causes the discontinuity in the stage–discharge relationship. The change from dunes to plane bed occurs at a larger discharge than the change from plane bed back to dunes.

The 1984 flood of the Meuse River reached a peak discharge of 2,231 m^3/s on February 12. A sequence of bathymetric profiles in the Bergsche Maas (lower Meuse) is given in Figure 8.15, where the amplitude and wavelength of bedforms are shown to change rapidly during floods. Soundings before the flood, $Q = 1,434$ m^3/s on February 8, were quite similar to those after the flood, $Q = 654$ m^3/s on February 20. At higher discharge, the large dunes showed rounded crests and some dunes measured up to 3 m in amplitude.

Comparing the results of February 8 and February 15, at a similar flow discharge, both the dune height and wavelength are smaller under rising discharge than under falling discharge. This loop-rating effect is expected because the time scale required for the formation of 2-m-high sand dunes is of the order of 1–3 days in the Bergsche Maas. Computer Problem 8.1 and Problem 8.7 provide instructive bedform calculation exercises with

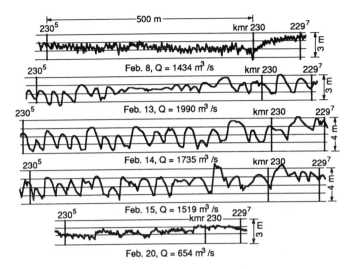

Figure 8.15. Dunes of the Bergsche Maas during the 1984 flood of the Meuse River (kmr denotes river kilometer; after Adriaanse, 1986)

field data. The following case study illustrates the changes in bedform configuration of the Rhine during a major flood.

Case Study 8.1 Rhine River, The Netherlands. During the 1988 flood, the discharge of the Rhine River at Lobith reached 10,274 m³/s on March 30 after a first peak of 8,324 m³/s on March 20, as shown in Figure CS8.1.1. Sequences of bathymetric profiles of the Rhine between km 863 and 866

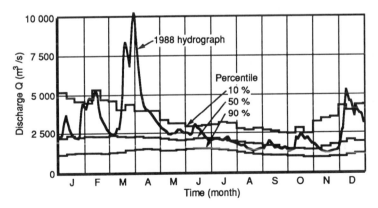

Figure CS8.1.1. Hydrograph of the Rhine River, 1988 (after Julien, 1992)

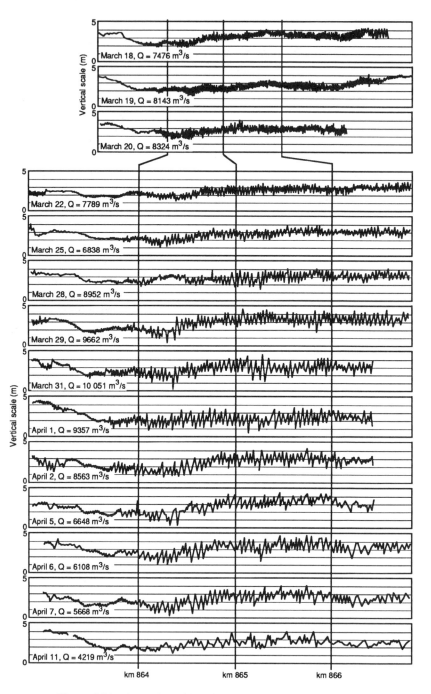

Figure CS8.1.2. Bed profiles of the Rhine Wall, 1988
(after Julien, 1992)

in The Netherlands are compiled in Figure CS8.1.2 for the period between March 18 and April 11, 1988. Dunes are shown to respond rapidly to changes in flow discharge. At high flow, dune crests are rounded, and irregular profiles develop in the falling limb of the hydrograph.

The bathymetric profiles of the Rhine River between km 865 and 866 were scrutinized to determine the average dune height Δ from the total bed elevation drop on the lee side of the dunes divided by the number of dunes over the 1-km reach. This average dune height roughly equals one-half to two-thirds of the maximum dune height Δ_m of the largest dune found over the 1-km reach. The average dune length Λ is calculated from the reach length divided by the number of dunes.

The dune height and the dune steepness increase with discharge. Looprating effects are not significant on daily records because the time scale for the formation of dunes in the Rhine is of the order of 6–12 hours. The corresponding dune height and steepness parameters in Figure CS8.1.3

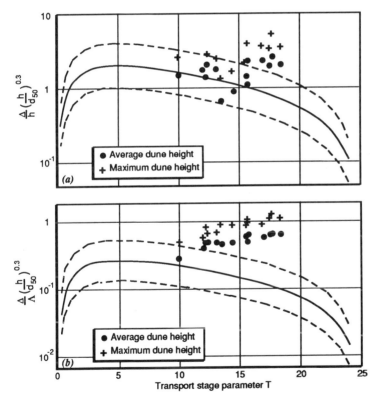

Figure CS8.1.3. Bedform height and steepness of the Rhine River, 1988 (after Julien, 1992)

show that the dune height parameter generally increases with the transport-stage parameter T, while the dune steepness parameter slightly increases with T.

Exercise

*8.1. Demonstrate that Equations (8.1) and (8.2) reduce to Equations (8.3) and (8.4).

*Problem 8.1

Identify the conditions under which the pressure and shear stress distributions vary linearly with flow depth.

Answer: Steady uniform flow.

**Problem 8.2

Determine the flow regime and type of bedform in the Rio Grande conveyance channel, given the mean velocity $V = 0.5$ m/s, the flow depth $h = 0.40$ m, the bed slope $S_0 = 52$ cm/km, and the grain size distributions $d_{50} = 0.24$ mm and $d_{65} = 0.35$ mm.

Problem 8.3

Check the type of bedform in a 200-ft-wide channel conveying 8500 ft^3/s in a channel sloping at 9.6×10^{-5} given the mean velocity $V = 3.6$ ft/s and the median grain diameter $d_m = 0.213$ mm.

Answer: $\tau_0 V = 0.25$ lb/ft $\times$ s, dunes from Simons and Richardson's (1966) diagram (Fig. 8.8).

*Problem 8.4

Predict the type and geometry of bedforms in a sand-bed channel, $d_{35} = 0.35$ mm and $d_{65} = 0.42$ mm, sloping at $S_0 = 0.001$ with flow depth $h = 1$ m when the water temperature is 40°F.

**Problem 8.5

A 20-m-wide alluvial channel conveys a discharge $Q = 45$ m^3/s. If the channel slope is $S_0 = 0.0003$ and the median sediment size is $d_m = 0.4$ mm, determine

(a) the flow depth from Engelund's method;
(b) the type of bedform; and
(c) the bedform geometry from van Rijn's method.

Problem 8.6

Dunes 6.4 m in height and 518 m in length were measured in the Mississippi River. If the flow depth is 38.7 m, the river energy slope 7.5 cm/km, the water temperature 16°C, the depth-averaged flow velocity 2.6 m/s, and the grain size $d_{50} = 0.25$ mm and $d_{90} = 0.59$ mm,

(a) check the type of bedform predicted by the methods of Liu, Chabert and Chauvin, Simons and Richardson, Bogardi, and van Rijn;
(b) estimate S'' from Equation (8.13); and
(c) calculate k_s from field data using Equation (8.15) and plot the results in Figure 8.13.

**Problem 8.7

Surveys of the Red Deer River indicate the formation of sand dunes 0.6 m high and 5.2 m long in bed material $d_{50} = 0.34$ mm and $d_{90} = 1.2$ mm. If the flow depth is 2 m, the average flow velocity 0.95 m/s, and the slope 7.4 cm/km,

(a) compare the bedform characteristics with those of Liu, Chabert and Chauvin, Simons and Richardson, Bogardi, Yalin, Engelund, and van Rijn;
(b) separate the total bed shear stress τ_b into grain and form shear stress components τ_b' and τ_b'' using the methods of Engelund and van Rijn; and
(c) compare the bedform length and height with the methods of Yalin and van Rijn.

*Computer Problem 8.1

From the Bergsche Maas bedform data given in the tabulation on the following page,

(a) calculate the grain Chézy coefficient C' and the sediment transport parameter T;
(b) plot the data on van Rijn's dune height and dune steepness diagrams; and
(c) calculate k_s from field measurements and plot in Figure 8.13.

Q (m^3/s)	S_f (cm/km)	h (m)	V (m/s)	d_s (μm)	Δ (m)	Λ (m)	T	C' $(m^{1/2}/s)$
2160.00	12.50	8.60	1.35	480	1.50	22.50	9.09	81.6
2160.00	12.50	8.00	1.35	410	1.00	14.00	9.05	84.2
2160.00	12.50	10.50	1.30	300	1.50	30.00		
2160.00	12.50	10.00	1.40	500	1.60	32.00		
2160.00	12.50	7.60	1.40	520	1.40	21.00		
2160.00	12.50	8.40	1.40	380	1.50	22.50		
2160.00	12.50	8.70	1.70	300	1.50	30.00		
2160.00	12.50	7.50	1.55	250	2.50	50.00		
2160.00	12.50	8.30	1.50	260	1.80	36.00		
2160.00	12.50	9.50	1.35	230	1.80	36.00		
2160.00	12.50	8.80	1.35	240	1.80	36.00		
2160.00	12.50	9.00	1.30	240	1.80	36.00		
2160.00	12.50	9.60	1.50	220	2.20	33.00		
2160.00	12.50	8.70	1.50	370	1.90	28.50		
2160.00	12.50	8.20	1.35	330	2.00	36.00		
2160.00	12.50	8.20	1.35	480	1.40	22.40		
2160.00	12.50	8.10	1.40	350	1.00	20.00		
2160.00	12.50	8.00	1.50	420	.60	9.00		
2160.00	12.50	7.80	1.50	410	.60	9.00		
2160.00	12.50	6.80	1.50	400	.40	8.00		
2160.00	12.50	6.40	1.50	270	.60	6.00		
2160.00	12.50	5.80	1.50	220	1.00	10.00		
2160.00	12.50	6.20	1.50	210	1.20	24.00		
2160.00	12.50	6.60	1.50	210	1.00	50.00		
2160.00	12.50	8.30	1.50	180	.90	36.00		
2160.00	12.50	8.10	1.35	400	1.50	22.50		

Source: After Julien (1992).

**Computer Problem 8.2

Consider bedforms and resistance to flow in the backwater profile analyzed in Computer Problem 3.1.

(a) Assume that the rigid boundary is replaced with uniform 1-mm sand. Select an appropriate bedform predictor and determine the type of bedform to be expected along the 25-km reach of the channel using previously calculated hydraulic parameters. (Check your results with a second bedform predictor.) Also determine the corresponding resistance to flow along the channel and recalculate the backwater profile. Briefly discuss the methods, assumptions, and results. Three sketches should be provided along the 25 km of the reach:

   (i) type of bedforms;

   (ii) Manning n or Darcy–Weisbach f; and

   (iii) water surface elevation.

(b) Repeat the procedure described in (a) after replacing the bed material with the sediment size distribution from Problem 2.4.

9

Bedload

Noncohesive bed particles enter motion as soon as the shear stress applied on the bed material exceeds the critical shear stress. Generally, silt and clay particles enter suspension (see Chapter 10), and sand and gravel particles roll and slide in a thin layer near the bed called the *bed layer*. Bedload, or contact load, refers to the transport of sediment particles that frequently maintain contact with the bed. Bedload transport can be treated either as a deterministic or as a probabilistic problem. Deterministic methods have been proposed by Duboys and by Meyer-Peter and Müller; probabilistic methods were developed by Kalinske and Einstein. Satisfactory estimates of bedload discharge are obtained from empirical methods presented in Section 9.1. Bed-material sampling is discussed in Section 9.2, and bed sediment discharge measurement techniques are summarized in Section 9.3.

The unit contact sediment discharge, or unit bedload discharge, refers to the flux of sediment per unit width per unit time that is in motion within the bed layer. The bed layer thickness is approximately twice the particle diameter, which is typically less than 1 mm in sand-bed channels and up to tens of centimeters in gravel-bed streams. Note that the bed layer thickness should not be mistaken for the laminar sublayer thickness defined in Chapter 6. The unit bedload discharge can be measured by weight q_{bw} with dimension M/T^3, by mass q_{bm} with dimension M/LT, or by volume q_{bv} with dimension L^2/T. Conversions from weight to mass or to volume involve the specific weight of sediment γ_s or the specific mass of sediment ρ_s, such that $q_{bw} = g q_{bm} = \rho_s g q_{bv} = \gamma_s q_{bv}$.

When integrated over the entire width of an open channel, the bedload discharge Q_b is obtained from $Q_b = \int_{\text{width}} q_b \, dw$. The dimensions of Q_{bw}, Q_{bm}, and Q_{bv} are ML/T^3, M/T, and L^3/T, respectively.

The bedload L_b transported in an open channel refers to the weight (mass or volume) of bed material conveyed within the bed layer. Time integration of the bedload discharge on a daily, monthly, or annual basis is obtained from $L_b = \int_{time} Q_b \, dt$. The fundamental dimensions of the load by weight, mass, and volume are ML/T^2, M, and L^3, respectively. It is most common to calculate sediment load by mass in metric tons (1,000 kg), by weight in pounds or tons (2,000 lb), by volume in cubic meters or cubic yards. It must be noted that only the volume of solids is considered in conversions, not the volume of loose material.

Sediment transport calculations by size fractions are obtained as

$$q_b = \Sigma \, \Delta p_i \, q_{bi} \tag{9.1}$$

where Δp_i is the fraction by weight of sediment particles of fraction i found in the bed, and q_{bi} is the unit sediment discharge of fraction i. Notice that $\Sigma_i \, \Delta p_i = 1$; calculations in percent are divided by 100.

9.1   Bedload equations

9.1.1   Duboys' equation

The pioneering contribution of Duboys (1879) is based on the concept that sediment moves in thin layers along the bed. The applied bed shear stress τ_0 must exceed the critical shear stress τ_c to initiate motion. The volume of bed material in motion per unit width and time q_{bv} in ft^2/s is calculated from

$$q_{bv} = \frac{0.173}{d_s^{3/4}} \tau_0 (\tau_0 - 0.0125 - 0.019 d_s) \tag{9.2}$$

where d_s is the particle size in mm and τ_0 is the boundary shear stress in lb/ft^2. Note that the critical shear stress ($\tau_c = 0.0125 + 0.019 d_s$; τ_c in lb/ft^2) is quite compatible with Figure 7.7 proposed almost a century after Duboys' contribution.

9.1.2   Meyer-Peter and Müller's equation

Meyer-Peter and Müller (1948) developed a complex bedload formula based on the median sediment size d_{50}. Chien (1956) demonstrated that the elaborate original formulation can be reduced to the following simple form:

$$\frac{q_{bv}}{\sqrt{(G-1)gd_s^3}} = 8(\tau_* - \tau_{*c})^{3/2} \tag{9.3a}$$

This formulation is most appropriate for channels with large width-depth ratios. The corresponding dimensional formulation for q_{bv} with dimensions of L^2/T is

$$q_{bv} \approx \frac{12.9}{\gamma_s \sqrt{\rho}} (\tau_0 - \tau_c)^{1.5} \qquad (9.3b)$$

The most complete formulation for composite channel configurations can be found in Simons and Senturk (1977) and in Richardson, Simons, and Julien (1990).

9.1.3   Einstein and Brown's equation

H. A. Einstein introduced the idea that grains move in steps proportional to their size. He defined the bed layer thickness as twice the particle diameter. He used extensive probability concepts to formulate a relationship for contact sediment discharge. The contact sediment discharge q_{bv} in volume of sediment per unit width and time (q_{bv} in L^2/T) is transformed, using Rubey's clear-water fall velocity ω_0 from Equation (5.23b), into a dimensionless volumetric unit sediment discharge q_{bv*} as

$$q_{bv*} = \frac{q_{bv}}{\omega_0 d_s}$$

$$= q_{bv} \left(\sqrt{(G-1)gd_s^3} \left\{ \sqrt{\frac{2}{3} + \frac{36\nu^2}{(G-1)gd_s^3}} - \sqrt{\frac{36\nu^2}{(G-1)gd_s^3}} \right\} \right)^{-1} \qquad (9.4)$$

The dimensionless rate of sediment transport q_{bv*} is shown in Figure 9.1 as a function of the Shields parameter $\tau_* = \tau_0/(\gamma_s - \gamma)d_s$, with measurements from Gilbert (1914), Meyer-Peter and Müller (1948), Bogardi (1974), and Wilson (1966). Brown (1950) suggested the following two relationships:

$$q_{bv*} = 2.15 e^{-0.391/\tau_*} \quad \text{when } \tau_* < 0.18 \qquad (9.5a)$$

and

$$q_{bv*} = 40\tau_*^3 \quad \text{when } 0.52 > \tau_* > 0.18 \qquad (9.5b)$$

Considering sediment transport data at high shear rates, $\tau_* > 0.52$, one obtains

$$q_{bv*} = 15\tau_*^{1.5} \quad \text{when } \tau_* > 0.52 \qquad (9.5c)$$

At such shear rates, however, sediment will also move in suspension, as discussed in Chapter 10.

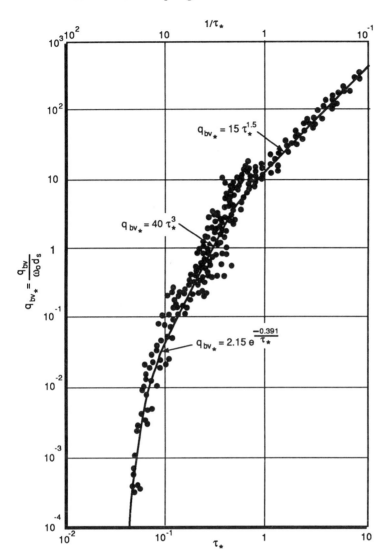

Figure 9.1. Dimensionless sediment discharge versus Shields parameter

9.2   Bed sediment sampling

Bed samples are usually collected at, or slightly below, the bed surface to define the particle size distribution and density of sediment particles available for transport. To obtain satisfactory submerged bed material samples, the samplers should enclose a volume of sediment and then isolate the sample from currents while the sampler is being lifted to the surface. The ease with which the sample can be transferred to a suitable container is also important.

Samplers for obtaining bed material are generally one of the following types: (1) drag bucket or scoop; (2) grab bucket or clamshell; (3) vertical pipe or core; (4) piston core; or (5) rotating bucket.

Shallow samples. With the drag bucket or scoop, some of the sample material may wash away, and the clamshell and drag bucket do not always close properly if the sample contains gravel or clays. Accordingly, when these samplers must be used, special control is required to ensure that the samples are representative of the bed material.

Vertical pipe or core samplers are essentially tubes that are forced into the streambed; the sample is retained inside the cylinder by creating a partial vacuum above the sample. Penetration in fine-grained sediment is easy, but penetration in sand usually is limited to about 0.5 m or less. These samplers generally yield good-quality samples and are inexpensive and simple to maintain.

The US BMH-53 is a piston core sampler (Fig. 9.2a) consisting of a 9-in.-long, 2-in.-diameter brass or stainless steel pipe with cutting edge and suction piston attached to a control rod. The piston is retracted as the cutting cylinder is forced into the streambed. The partial vacuum in the sampling chamber, which develops as the piston is withdrawn, is of assistance in collecting and holding the sample in the cylinder. This sampler can be used only in streams that are shallow enough to be waded in.

Deep samples. The US BM-54, US BM-60, and Shipek are rotating bucket samplers. The US BM-54 shown in Figure 9.2b weighs 100 lb; it is designed to be suspended from a cable and to scoop up a sample of the bed sediment that is 3 in. in width and about 2 in. in maximum depth. When the sampler contacts the streambed with the bucket completely retracted, the tension in the suspension cable is released and a heavy coil spring quickly rotates the bucket through 180° to scoop up the sample. A rubber stop prevents any sediment from being lost.

The US BMH-60 bed material sampler shown in Figure 9.2c is similar to the US BM-54 and was developed for both handline and cable suspension. The sampler weighs 30 lb if made of aluminum, and 40 lb if made of brass. It is used to collect samples in streams with low velocities but with depths beyond the range of the US BMH-53 sampler.

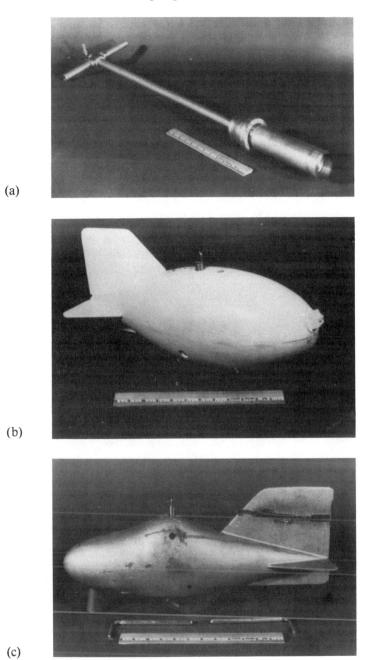

(a)

(b)

(c)

Figure 9.2. Bed sediment samplers (from Edwards and Glysson, 1988)

Coarse particle samples. Materials coarser than gravel and cobble are extremely difficult to sample effectively because penetration is difficult and large quantities of material must be collected. Strictly speaking, hundreds to thousands of pounds of bed material are required for an accurate determination of the particle size distribution. Manual collection and measurement are necessary to determine a representative sample of such material.

The recommended method for wadable streams is to use a grid pattern to locate sampling points. The particle at each grid point is retrieved, and its intermediate diameter measured and recorded. Where a grid point is over sand or finer material, a small volume (about 15 ml) is collected and combined with samples from other such points for sieve analysis. At least 200 points should be sampled in order for the relative quantity of some of the coarser sizes to be accurate.

9.3   Bedload measurements

Direct measurements of bedload discharge are so difficult that no standard procedure is available in spite of the intensive research efforts devoted to this problem. Bedload measurements are usually possible with samplers or other devices, including sediment traps, bedload samplers, and vortex tubes, or other techniques, such as tracer techniques or measurement of the migration of bedforms.

Box and basket samplers and sediment traps can sometimes be installed in small streams at a reasonable cost. These direct measuring devices measure the volume of bed material in motion near the bed during major events. The total volume or weight of sediment accumulated in the trap can be determined after each event, although very little information on the rate of sediment transport at a given discharge can be obtained.

Tracer techniques can be applied in coarse bed material streams by painting, staining, or radio tracking coarse particles from the bed. The position of the particles after a major event indicates the distance traveled during the flood and reflects sediment transport. The use of radioactive tracers is discouraged, however, for environmental reasons.

In sand-bed channels, the rate of bedload transport depends largely on the motion of large bedforms such as dunes. Since large sediment volumes are contained within dunes, their motion can be monitored and the bedload discharge corresponds to the dune volume divided by the time required for a full wavelength migration.

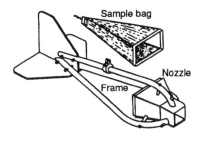

Figure 9.3. Helly-Smith bedload sampler (from Emmett, 1979)

Bedload samplers are most useful for providing the sediment size distribution and qualitative information on the rates of sediment transport in the layer extending 0.3 ft above the bed. The efficiency of bedload samplers such as the Helley-Smith, sketched in Figure 9.3, depends on the size fractions: Very coarse material will undoubtedly not enter the sampler, and very fine material will be washed through the sample bag.

Sampling over a long period of time may cause clogging of the sample bag and bias the measurements. Also, when the sampler is positioned over gravel- and cobble-bed streams, substantial amounts of finer particles (sand particles) will be transported underneath it. The Helley-Smith sampler seems best suited to coarse sand and fine gravel-bed streams, given the primary advantages of low cost and great mobility.

Another type of sampler is the vortex tube (Fig. 9.4), which has proved to be extremely effective in the removal of bedload in narrow open channels. Some vortex tubes have been effective in removing coarse bed material up to gravel and cobble size in laboratories, irrigation canals, and mountain streams. The main feature of the vortex tube is a vented circular tube with an opening along the top side mounted flush with the bed elevation. As water flows over the tube, the shearing action across the opening sets up vortex motion within and along the tube. This whirling action pulls the sediment particles passing over the lip of the opening and carries the particles to the outlet of the vortex tube. Case Study 9.1 provides detailed bedload calculations using the methods described in Section 9.1.

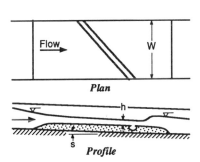

Figure 9.4. Vortex tube sediment ejector

Case Study 9.1   Mountain Creek, United States. Mountain Creek near Greenville, South Carolina, is a small, 14-ft-wide sand-bed stream. The geometric mean sediment size is 0.86 mm with standard deviation $\sigma_g = 1.8$, $d_{35} = 0.68$ mm, $d_{50} = 0.86$ mm, $d_{65} = 1.08$ mm, and $d_{90} = 1.88$ mm. The complete sieve analysis and sediment-rating measurements are given in the following tabulation:

Particle size distribution		Sediment-rating measurements		
Sieve opening (mm)	% Finer	Flow depth (ft)	Unit discharge (ft^2/s)	q_{bw} $(\text{lb}/\text{ft} \times \text{s})$
0.074	0.07	0.16	0.21	—
0.125	0.33	0.18	0.23	—
0.246	1.70	0.22	0.32	0.004
0.351	6.20	0.25	0.40	0.006
0.495	19.00	0.27	0.48	0.007
0.701	37.30	0.29	0.52	0.009
0.991	60.50	0.32	0.60	0.012
1.400	79.40	0.40	0.87	0.030
1.980	90.40	0.43	1.10	0.039
3.960	99.30	0.60	1.50	—

The water surface slope of Mountain Creek varied between 0.00155 and 0.0016 during the measurements. The unit discharge increased from 0.2 to 1.1 ft^2/s at flow depths corresponding to 0.16–0.43 ft, as indicated in the tabulation. Assuming water temperature at 78°F, calculate bedload from the methods of Duboys, Meyer-Peter and Müller, and Einstein and Brown using the median grain size, and compare calculations by size fractions with field measurements.

Calculations based on the median grain size. Bedload calculations for $d_{50} = 0.86$ mm $= 0.00282$ ft, $S_f = S_0 = 0.0016$, $h = 0.6$ ft, $\tau_0 = \gamma h S_f = 0.06$ lb/ft^2, $T° = 78°$F, and $\nu = 1 \times 10^{-5}$ ft^2/s follow:

  (a) Duboys' equation:

$$q_{bv} = \frac{0.173}{d_s^{3/4}} \tau_0 (\tau_0 - 0.0125 - 0.019 d_s)$$

$$= \frac{0.173}{(0.86)^{3/4}} 0.06(0.06 - 0.0125 - 0.019 \times 0.86)$$

$$= 3.6 \times 10^{-4} \text{ ft}^2/\text{s}$$

$$q_{bw} = \gamma_s q_{bv} = 0.0595 \text{ lb/ft} \times \text{s}$$

(b) Meyer-Peter and Müller equation with $\tau_{*c} = 0.047$:

$$\tau_* = \frac{\tau_0}{(\gamma_s - \gamma)d_s} = \frac{0.06 \text{ lb} \times \text{ft}^3}{\text{ft}^2 \times 1.65 \times 62.4 \text{ lb} \times 0.00282 \text{ ft}} = 0.206$$

$$q_{bv} = \sqrt{(G-1)gd_s^3} \, 8(\tau_* - 0.047)^{3/2}$$

$$= \sqrt{1.65 \times \frac{32.2 \text{ ft}}{\text{s}^2} \times (0.00282)^3 \text{ ft}^3} \times 8 \times (0.206 - 0.047)^{3/2}$$

$$= 5.5 \times 10^{-4} \text{ ft}^2/\text{s}$$

$$q_{bw} = \gamma_s q_{bv} = 0.091 \text{ lb/ft} \times \text{s}$$

(c) Einstein–Brown equation:

$$\tau_* = 0.206 > 0.18$$

$$q_{bv*} = 40\tau_*^3 = 40(0.206)^3 = 0.35$$

$$X_e = \frac{36\nu_m^2}{(G-1)gd_s^3} = \frac{36 \times 10^{-10} \text{ ft}^4\text{s}^2}{\text{s}^2 (1.65) \times 32.2 \text{ ft} \times (0.00282)^3 \text{ ft}^3} = 0.003$$

$$q_{bv} = q_{bv*}\sqrt{(G-1)gd_s^3} \left[\sqrt{\tfrac{2}{3} + 0.003} - \sqrt{0.003} \right]$$

$$= 0.35 \sqrt{1.65 \times \frac{32.2 \text{ ft}}{\text{s}^2}(0.00282)^3 \text{ ft}^3} \times 0.763$$

$$= 2.91 \times 10^{-4} \text{ ft}^2/\text{s}$$

$$q_{bw} = \gamma_s q_{bv} = 0.048 \text{ lb/ft} \times \text{s}$$

Calculations by size fractions. The weight fraction Δp_i for each size fraction is first determined from the sediment size distribution. Calculations by size fractions for each method at different flow depths are summarized in the following tables. The calculated results are compared with field measurements in Figure CS9.1.1.

d_s (mm)	Δp_i	Duboys $q_{bvi} \Delta p_i$ (10^{-6} ft²/s)	Meyer-Peter and Müller $q_{bvi} \Delta p_i$ (10^{-6} ft²/s)	Einstein– Brown $q_{bvi} \Delta p_i$ (10^{-6} ft²/s)
$h = 0.2$ ft, $\tau_0 = \gamma_m hS = 0.02$ lb/ft²				
0.074	0.00200	0.295	0.28	0.410
0.125	0.00815	0.682	1.10	1.100
0.246	0.02935	0.810	3.30	1.900
0.351	0.08650	0.520	8.30	3.900

d_s (mm)	Δp_i	Duboys $q_{bvi}\,\Delta p_i$ $(10^{-6}\ \text{ft}^2/\text{s})$	Meyer-Peter and Müller $q_{bvi}\,\Delta p_i$ $(10^{-6}\ \text{ft}^2/\text{s})$	Einstein-Brown $q_{bvi}\,\Delta p_i$ $(10^{-6}\ \text{ft}^2/\text{s})$
0.495	0.15550	0	11.60	4.700
0.700	0.20750	0	9.60	2.700
0.990	0.21050	0	3.20	0.700
1.400	0.14950	0	0	0.050
1.980	0.09950	0	0	0.001
3.960	0.05150	0	0	0
Total	$q_{bv} =$	$2.31 \times 10^{-6}\ \text{ft}^2/\text{s}$	$37.5 \times 10^{-6}\ \text{ft}^2/\text{s}$	$15.6 \times 10^{-6}\ \text{ft}^2/\text{s}$
	$q_{bw} =$	$0.0004\ \text{lb/ft} \times \text{s}$	$0.0062\ \text{lb/ft} \times \text{s}$	$0.0026\ \text{lb/ft} \times \text{s}$

$h = 0.6$ ft, $\tau_0 = 0.06$ lb/ft^2

d_s (mm)	Δp_i	Duboys	Meyer-Peter and Müller	Einstein-Brown
0.074	0.00200	6.7	1.6	11.2
0.125	0.00815	18.1	6.3	31.2
0.246	0.02935	37.2	21.7	52.0
0.351	0.08650	80.1	61.1	95.4
0.495	0.15550	103.7	103.0	106.0
0.700	0.20750	95.8	124.8	85.8
0.990	0.21050	62.8	109.1	55.2
1.400	0.14950	25.0	61.0	26.9
1.980	0.09950	6.0	26.7	8.4
3.960	0.05150	0	0	0.16
Total	$q_{bv} =$	$435.6 \times 10^{-6}\ \text{ft}^2/\text{s}$	$515.5 \times 10^{-6}\ \text{ft}^2/\text{s}$	$472.3 \times 10^{-6}\ \text{ft}^2/\text{s}$
	$q_{bw} =$	$0.072\ \text{lb/ft} \times \text{s}$	$0.085\ \text{lb/ft} \times \text{s}$	$0.078\ \text{lb/ft} \times \text{s}$

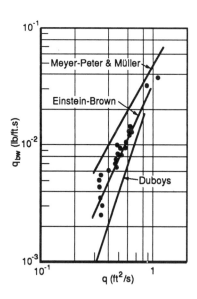

Figure CS9.1.1. Sediment-rating curve for Mountain Creek

*Problem 9.1

Calculate the unit bedload discharge for a channel given the slope $S_0 = 0.01$, the flow depth $h = 20$ cm, and the grain size $d_{50} = 15$ mm. From Duboys' equation, calculate q_{bw} in lb/ft × s, and q_{bv} in ft^2/s.

Answer: $q_{bw} = 0.17$ lb/ft × s; $q_{bv} = 1.03 \times 10^{-3}$ ft^2/s.

*Problem 9.2

Use Meyer-Peter and Müller's method to calculate q_{bm} in kg/m × s and q_{bv} in m^2/s for the conditions given in Problem 9.1.

Problem 9.3

Use Einstein and Brown's method to calculate the bedload transport rate in a 100-m-wide coarse sand-bed channel with slope $S_0 = 0.003$ when the applied shear stress τ_0 equals τ_c. Determine the transport rate Q_{bv} in m^3/s and in ft^3/s.

Answer: $Q_{bv} = 3.27 \times 10^{-8}$ m^3/s $= 1.15 \times 10^{-6}$ ft^3/s.

*Problem 9.4

Use Einstein and Brown's bedload equation to calculate q_{bm} in kg/m × s and q_{bw} in lb/ft × s for the conditions given in Problem 9.1.

**Problem 9.5

Use the three methods detailed in this chapter to calculate the daily bedload in metric tons in a 20-m-wide medium gravel-bed canal with a slope $S_0 = 0.001$ and at a flow depth $h = 2$ m. Compare the results in metric tons per day.

Answer: Duboys: $Q_{bm} = 791$ tons/day; Meyer-Peter and Müller: 2,431 tons/day; Einstein and Brown: 884 tons/day.

*Problem 9.6

(a) Which bed sediment sampler would you recommend for the canal in Problem 9.5?

(b) Which method would you recommend for measuring and controlling bedload in the same canal?

*Problem 9.7

With reference to the Bergsche Maas bedform data given in Computer Problem 8.1,

(a) calculate the bedload sediment transport; and
(b) estimate the time required for the bedload to fill the volume of a representative dune.

**Computer Problem 9.1

Consider the channel reach analyzed in Computer Problems 3.1 and 8.2. Select an appropriate bedload relationship to calculate the bed sediment discharge in tons/m × day for the uniform 1-mm sand in Computer Problem 8.2. Plot the results along the 25-km reach and discuss the method, assumptions, and results.

*Computer Problem 9.2

Write a computer program to calculate the bedload transport rate by size fraction from the methods of Duboys, Meyer-Peter and Müller, and Einstein and Brown, and repeat the calculations of the tabulation in Case Study 9.1 at $h = 0.4$ ft and $\tau_0 = 0.04$ lb/ft^2.

Answer

Duboys: $q_{bv} = 122.3 \times 10^{-6}$ ft^2/s, $q_{bw} = 0.0202$ lb/ft × s;
Meyer-Peter and Müller: $q_{bv} = 221 \times 10^{-6}$ ft^2/s, $q_{bw} = 0.0365$ lb/ft × s;
Einstein and Brown: $q_{bv} = 140.9 \times 10^{-6}$ ft^2/s, $q_{bw} = 0.0233$ lb/ft × s.

10

Suspended load

As the hydraulic forces exerted on sediment particles exceed the threshold condition for beginning of motion, coarse sediment particles move in contact with the bed surface, as described in Chapter 9. Finer particles are brought into suspension when turbulent velocity fluctuations are sufficiently large to maintain the particles within the mass of fluid without frequent bed contact.

This chapter examines the concentration of sediment particles held in suspension (Section 10.1). The governing equations of turbulent diffusion are presented in Section 10.2 followed by turbulent mixing of washload in Section 10.3. Equilibrium vertical concentration profiles (Section 10.4) support the analysis of suspended load in Section 10.5 and hyperconcentrations in Section 10.6. Field measurement techniques are covered in Section 10.7. Five examples illustrate the computation procedures.

10.1 Sediment concentration

To avoid misinterpretation, the term *concentration* requires clarification. The units used in the measurement of sediment concentration vary with the range of concentrations and the standard measurement techniques utilized in different countries. The most common unit for sediment concentration is milligrams per liter, which describes the ratio of the mass of sediment particles to the volume of the water–sediment mixture. Other units include kilograms per cubic meter ($1 \text{ mg/l} = 1 \text{ g/m}^3$), the volumetric sediment concentration C_v, the concentration in parts per million C_{ppm}, and the concentration by weight C_w:

$$C_v = \frac{\text{sediment volume}}{\text{total volume}} = \frac{\forall_s}{\forall_t} = 1 - p_0 \qquad (10.1a)$$

$$C_w = \frac{\text{sediment weight}}{\text{total weight}} = \frac{C_v G}{1 + (G-1)C_v} \qquad (10.1b)$$

Table 10.1. *Equivalent concentrations for C_v, C_w, C_{ppm}, and $C_{mg/l}$*

C_v		C_w	C_{ppm}	$C_{mg/l}$
Suspension				
0.001		0.00264	2,645	2,650
0.0025		0.00659	6,598	6,625
0.005		0.01314	13,141	13,250
0.0075		0.01963	19,632	19,875
0.01		0.02607	26,069	26,500
0.025		0.06363	63,625	66,250
Hyperconcentration				
0.05		0.12240	122,401	132,500
0.075		0.17686	176,863	198,750
0.1		0.22747	227,467	265,000
0.25		0.46903	469,027	662,500
0.5		0.72603	726,027	1,325,000
0.75		0.88827	888,268	1,987,500

Note: Calculations are based on mean density of water of 1 g/ml and specific gravity of sediment $G = 2.65$.

in which $G = \gamma_s/\gamma$, and

$$C_{ppm} = 10^6 C_w \tag{10.1c}$$

Note that the percentage by weight C_{ppm} is given by 1,000,000 times the weight of sediment over the weight of the water–sediment mixture. The corresponding concentration in milligrams per liter is then calculated by the following formula:

$$C_{mg/l} = \frac{1\,\text{mg/l}\,GC_{ppm}}{G + (1 - G)10^{-6}C_{ppm}} = 10^6\,\text{mg/l}\,GC_v \tag{10.1d}$$

The conversion factors in going from C_{ppm} to $C_{mg/l}$ are given in Table 10.1. Notice that there is less than 10% difference between C_{ppm} and $C_{mg/l}$, at concentrations $C_{ppm} < 145,000$.

In the laboratory, the sediment concentration $C_{mg/l}$ is measured as 1,000,000 times the ratio of the dry mass of sediment in grams to the volume of the water–sediment mixture in cubic centimeters (1 cm^3 = 1 ml). Two methods are commonly used: evaporation and filtration. The evaporation method is employed when the sediment concentration of samples exceeds 2,000–10,000 mg/l; the filtration method is preferred at lower concentrations. The lower limit applies when the sample consists mostly of fine material (silt and clay), and the upper limit when the sample is mostly sand. For samples having low sediment concentration, the evaporation method requires a correction if the dissolved solids content is high.

Since the concentration C varies with space (x, y, z) and time (t), several average values can be considered:

1. Time-averaged concentration C_t integrated over the sampling time t_s:

$$C_t(x, y, z, t_0) = \frac{1}{t_s} \int_{t_0}^{t_0 + t_s} C(x, y, z, t) \, dt \qquad (10.2a)$$

2. Spatial-averaged concentration $C_\forall$ integrated over a volume $\forall$:

$$C_\forall(x_0, y_0, z_0, t) = \frac{1}{\forall} \int_\forall C(x, y, z, t) \, d\forall \qquad (10.2b)$$

3. Flux-averaged concentration C_f, which, when multiplied by the total flow discharge Q, gives the exact advective mass flux passing through a given cross section A; thus,

$$C_f = \frac{1}{Q} \int_A C v_x \, dA \qquad (10.2c)$$

The flux-averaged concentration C_f is used when defining the concentration of sediments at a given stream cross section.

10.2   Advection–diffusion equation

The equation governing the conservation of mass can be applied to a small cubic control volume to derive the sediment continuity relationship. The rate of mass change per unit volume, $\partial C / \partial t$, simply equals the difference of mass fluxes across the faces of the control volume. The derivation previously detailed in Example 3.1 states that the sediment continuity relationship can be written as

$$\frac{\partial C}{\partial t} + \frac{\partial \hat{q}_{tx}}{\partial x} + \frac{\partial \hat{q}_{ty}}{\partial y} + \frac{\partial \hat{q}_{tz}}{\partial z} = \dot{C} \qquad (10.3)$$

in which C is the spatial-averaged sediment concentration inside the infinitesimal control volume, $\hat{q}_{tx}$, $\hat{q}_{ty}$, $\hat{q}_{tz}$ are the mass fluxes through the faces of the control volume, and $\dot{C}$ is the rate of sediment per unit volume supplied from an external source. Note that the units of C are not important as long as they are consistent with those of $\hat{q}_t$ and $\dot{C}$.

Expanding the derivation in Example 3.1, one recognizes three types of mass fluxes across the faces of the control volume: advective fluxes, diffusive fluxes, and mixing fluxes. These can be written in a simple mathematical form as

$$\hat{q}_{tx} = v_x C - (D + \epsilon_x)\frac{\partial C}{\partial x} \tag{10.4a}$$

$$\hat{q}_{ty} = v_y C - (D + \epsilon_y)\frac{\partial C}{\partial y} \tag{10.4b}$$

$$\hat{q}_{tz} = \underbrace{v_z C}_{\substack{\text{advective} \\ \text{fluxes}}} - \underbrace{(D + \epsilon_z)\frac{\partial C}{\partial z}}_{\substack{\text{diffusive and} \\ \text{mixing fluxes}}} \tag{10.4c}$$

The advective fluxes describe the transport of sediments imparted by velocity currents. The rate of mass transport per unit area carried by advection is obtained from the product of sediment concentration and the velocity components v_x, v_y, and v_z, respectively (e.g., Example 3.1).

Two additional fluxes must be considered in sediment transport. Molecular diffusion refers to the scattering of sediment particles by random molecular motion, as described by Fick's law. Turbulent diffusion induces the scattering of sediment particles due to turbulent fluid motion. The rate of sediment transport for both fluxes is proportional to the concentration gradient. For diffusion in laminar flow, the proportionality constant D is called the molecular diffusion coefficient and has dimensions of L^2/T. The minus sign in Equations (10.4a)–(10.4c) indicates that the mass flux is directed toward the direction of decreasing concentration (or negative concentration gradient). For mixing in turbulent flow, the turbulent mixing coefficients ϵ_x, ϵ_y, and ϵ_z determine the magnitude of turbulent diffusion. Mass transport in turbulent flow is also proportional to the concentration gradient. The molecular diffusion coefficient D is several orders of magnitude smaller than the turbulent mixing coefficients ϵ in turbulent flow.

The general relationship describing conservation of mass for incompressible dilute sediment suspensions subjected to diffusion, mixing, and advection (constant D, ϵ_x, ϵ_y, ϵ_z) with point sediment sources is obtained after substituting Equation (10.4) into Equation (10.3):

$$\frac{\partial C}{\partial t} + \frac{\partial v_x C}{\partial x} + \frac{\partial v_y C}{\partial y} + \frac{\partial v_z C}{\partial z}$$

$$= \dot{C} + (D + \epsilon_x)\frac{\partial^2 C}{\partial x^2} + (D + \epsilon_y)\frac{\partial^2 C}{\partial y^2} + (D + \epsilon_z)\frac{\partial^2 C}{\partial z^2} \tag{10.5a}$$

As opposed to the Navier–Stokes equations discussed in Chapter 5, Equation (10.5a) can be easily solved owing to the linearity of the independent parameter C. This equation has numerous applications in the field of sediment and contaminant transport in open channels.

Owing to the conservation of fluid mass [Equation (3.10)] for an incompressible fluid at a low sediment concentration, the advection–diffusion relationship can be rewritten as

$$\underbrace{\frac{\partial C}{\partial t}}_{\text{mass change}} + \underbrace{v_x\frac{\partial C}{\partial x} + v_y\frac{\partial C}{\partial y} + v_z\frac{\partial C}{\partial z}}_{\text{advective terms}}$$

$$= \underbrace{\dot{C}}_{\text{source}} + \underbrace{(D+\epsilon_x)\frac{\partial^2 C}{\partial x^2} + (D+\epsilon_y)\frac{\partial^2 C}{\partial y^2} + (D+\epsilon_z)\frac{\partial^2 C}{\partial z^z}}_{\text{diffusive and mixing terms}} \qquad (10.5b)$$

In laminar flow, the turbulent mixing coefficients vanish ($\epsilon_x = \epsilon_y = \epsilon_z = 0$). Conversely, in turbulent flows, the molecular diffusion coefficient D is negligible compared with the turbulent mixing coefficients ($D \ll \epsilon$).

10.3 Turbulent mixing of washload

The concept of washload refers to fine particles that are easily washed away by the flow – more specifically, particles in transport that are not found in significant amounts in the bed material. The limiting sediment size between washload and bed sediment load corresponds to the point at which the sediment transport capacity equals the sediment supply from upstream. Einstein suggested that the largest sediment size of washload may be arbitrarily chosen as the grain diameter d_{10} of which 10% by weight of the bed sediment is finer. Washload does not have any direct relation to sediment size, although washload often corresponds to the silts and clays in sand-bed channels.

Three criteria have been recommended for separating the washload from bed sediment load: transport capacity and supply curves, grain diameter d_{10}, and sediment size finer than 0.0625 mm. Some difficulties in the use of these criteria have been experienced in streams with large concentrations of fines and in gravel- and cobble-bed streams.

10.3.1 One-dimensional mixing

Consider the case of one-dimensional turbulent mixing in a channel transverse y direction, $\epsilon_y = \epsilon_t$, of a mass m of fine sediment without settling (washload) introduced at time zero at the y origin. The governing one-dimensional equation [Eq. (10.5)] for turbulent mixing, given $\partial C/\partial x = \partial C/\partial z = v_y = \partial^2 C/\partial x^2 = \partial^2 C/\partial z^2 = 0$, reduces to

$$\frac{\partial C}{\partial t} = \frac{\epsilon_t \partial^2 C}{\partial y^2} \tag{10.6}$$

The solution to this equation gives the concentration $C(y, t)$ in space and time,

$$C(y, t) = \frac{m}{A\sqrt{4\pi\epsilon_t t}} e^{-y^2/4\epsilon_t t} \tag{10.7}$$

in which the mass $m = \int_\forall C_{mg/l} \, d\forall$ is the volume integral of the concentration given the cross-sectional area A in the x-z plane. The properties of this normal distribution are that the variance increases linearly with time, $\sigma_t^2 = 2\epsilon_t t$, and a practical estimate of the width of the mixing cloud is simply obtained from $4\sigma_t = 4\sqrt{2\epsilon_t t}$.

10.3.2   Mixing coefficients, length and time scales

Consider a straight rectangular channel of width W, flow depth h, and shear velocity u_*. Turbulent mixing and dispersion are described by coefficients bearing the fundamental dimensions L^2/T. The vertical mixing coefficient ϵ_v, the transversal mixing coefficient ϵ_t, and the longitudinal dispersion coefficient K_d are empirical functions of the product hu_* as

$$\epsilon_v \cong 0.067hu_* \tag{10.8a}$$

$$\epsilon_t \cong 0.15hu_* \quad \text{in straight channels} \tag{10.8b}$$

$$\epsilon_t \cong 0.6hu_* \quad \text{in natural channels} \tag{10.8c}$$

$$K_d \cong 250hu_* \quad \text{or} \quad 0.011\frac{V^2W^2}{hu_*} \tag{10.8d}$$

It is observed that the vertical, transversal, and dispersion coefficients increase by orders of magnitude, respectively. Given the property of the variance for normal distributions in Section 10.3.1, the vertical time scale t_v, the transversal time scale t_t, and the longitudinal dispersion time scale t_d in a stream are defined after substituting t_t, respectively, with the average flow depth h ($h^2 = 2\epsilon_v t_v$), the channel width W ($W^2 = 2\epsilon_t t_t$), and the dispersion length X_d ($X_d^2 = 2K_d t_d$):

$$t_v = \frac{X_v}{V} \cong \frac{h}{0.1u_*} \tag{10.9a}$$

$$t_t = \frac{X_t}{V} \cong \frac{W^2}{hu_*} \tag{10.9b}$$

$$t_d = \frac{X_d}{V} \cong \frac{X_d^2}{500hu_*} \quad \text{where } t_d > t_t \tag{10.9c}$$

Traveling downstream at the mean flow velocity V, the corresponding length scales for complete vertical mixing X_v, complete transversal mixing X_t, and longitudinal dispersion X_d are given by

$$X_v = \frac{hV}{0.1u_*} \tag{10.10a}$$

$$X_t = \frac{VW^2}{hu_*} \tag{10.10b}$$

and

$$X_d = \frac{500hu_*}{V} \tag{10.10c}$$

First-order approximations of the distances X_v and X_t required for vertical and transversal mixing indicate that $t_t/t_v = X_t/X_v = 0.1W^2/h^2$. It is concluded that vertical mixing occurs before transversal mixing unless $W < 3h$. Example 10.1 provides calculations of time and length scales.

Example 10.1   Application to mixing time and length scales. Determine an approximate time scale and the downstream distance required for complete vertical and transversal mixing of washload in a gently meandering 600-ft-wide stream. The stream is 30 ft deep. The average flow velocity is $V = 2$ ft/s and the shear velocity is 0.2 ft/s.

Step 1. Vertical mixing occurs within a time t_v approximated by Equation (10.9a):

$$t_v \cong \frac{h}{0.1u_*} = \frac{30 \text{ ft} \times \text{s}}{0.1 \times 0.2 \text{ ft}} = 1,500 \text{ s} = 25 \text{ min}$$

The corresponding downstream distance is $X_v = Vt_v = 2.0$ ft/s $\times 1,500$ s $= 3,000$ ft. Notice that complete vertical mixing occurs at a downstream distance roughly equal to 100 times the flow depth.

Step 2. Transversal mixing should be complete at a time t_t approximated by Equation (10.9b):

$$t_t \cong \frac{W^2}{hu_*} = \frac{(600)^2 \text{ ft}^2 \times \text{s}}{30 \text{ ft} \times 0.2 \text{ ft}} = 60,000 \text{ s} \cong 17 \text{ h}$$

The corresponding downstream distance X_t is

$$X_t = Vt_t = \frac{2.0 \text{ ft}}{\text{s}} \times 60,000 \text{ s} = 120,000 \text{ ft} \cong 23 \text{ mi}$$

Vertical mixing is completed long before transversal mixing.

10.3.3   Mixing from steady point sources

Consider the steady supply of washload as a centerline point source of concentration C_0 at a mass rate $\dot{m}$ into a stream of width W, depth h, and average flow velocity V, as sketched in Figure 10.1a. Assuming that advection is dominant in the downstream x direction while mixing occurs in the transversal y direction, $\epsilon_y = \epsilon_t$, the governing equation [Eq. (10.5)] reduces to

$$\frac{\partial C}{\partial t} + \frac{V \partial C}{\partial x} = \epsilon_t \frac{\partial^2 C}{\partial y^2} \qquad (10.11)$$

The solution, after a coordinate transformation, $x' = x - Vt$, becomes similar to Equation (10.7), in which x is replaced by x'. Advection and

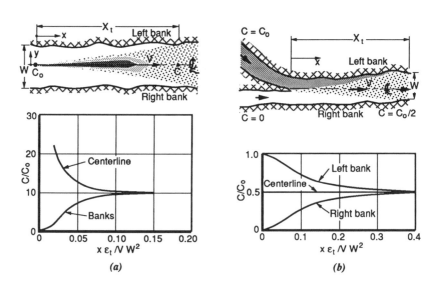

(a)   (b)

Figure 10.1. (a) Plume for steady point source; (b) mixing at stream confluence (after Fischer et al., 1979)

mixing are separate and additive processes. In other words, advective mixing is the same as mixing in a stagnant fluid when viewed in a coordinate system moving at speed V. Assuming that complete vertical mixing occurs long before transversal mixing, the approximate solution when $t \gg 2\epsilon_y/V^2$ in an infinitely wide channel is

$$C(x,y) = \frac{\dot{m}}{h\sqrt{4\pi\epsilon_t xV}} e^{-y^2 V/4\epsilon_t x} \tag{10.12}$$

In channels of finite width W, the relative concentration C/C_0 with $C_0 = \dot{m}/hVW$ for a midstream point source is plotted in Figure 10.1a. A reasonable length X_t for complete transversal mixing is $X_t \cong 0.15VW^2/\epsilon_t$ for centerline injection. In the case of side injection, the solution can be found by analogy with a double channel width. Thus, the distance X_t for complete transversal mixing is $X_t \cong 0.6VW^2/\epsilon_t$ for side injection of fine sediment such as washload from bank erosion.

The case of mixing of two streams of equal discharge is shown in Figure 10.1b. A reasonable length for complete transversal mixing is $X_t \cong 0.4VW^2/\epsilon_t$. Example 10.2 presents calculations of turbulent mixing of a steady point source of washload.

Example 10.2   Application to resuspension from a dredge. A dredge causes the steady resuspension of 3 million gallons per day of clay at a 200,000-ppm concentration near the centerline of a 30-ft-deep and 1,300-ft-wide river. If the mean flow velocity is 2 ft/s and the shear velocity is 0.2 ft/s, determine the width of the plume and the maximum concentration 1,000 ft downstream from the dredge (assume a uniform vertical distribution).

1 million gal/day = 0.04382 m³/s

Step 1. The mass flux is

$$\dot{m} = QC = 3 \times 0.04382 \text{ m}^3/\text{s} \times 35.32 \text{ ft}^3/\text{m}^3 \times 2 \times 10^5 \text{ ppm}$$
$$= 9.3 \times 10^5 \text{ ppm} \times \text{ft}^3/\text{s}$$

Step 2. The transverse mixing coefficient $\epsilon_t \cong 0.6hu_*$ is

$$\epsilon_t = 0.6 \times 30 \text{ ft} \times 0.2 \text{ ft/s} = 3.6 \text{ ft}^2/\text{s}$$

Step 3. The plume width $4\sigma_t$ is approximately

$$4\sigma_t \cong 4\sqrt{\frac{2\epsilon_t x}{V}} = 4\sqrt{2 \times \frac{3.6 \text{ ft}^2}{\text{s}} \times \frac{1,000 \text{ ft} \times \text{s}}{2 \text{ ft}}}$$

$$4\sigma_t \cong 240 \text{ ft}$$

Step 4. The maximum centerline concentration from Equation (10.12), at $x = 1,000$ ft and $y = 0$, is

$$C_{max} = \frac{\dot{m}}{h\sqrt{4\pi\epsilon_t x V}} = \frac{9.30 \times 10^5 \text{ ft}^3 \times \text{ppm} \times 1}{\text{s} \times 30 \text{ ft} \sqrt{4\pi \times 3.6 \text{ ft}^2/\text{s} \times 1,000 \text{ ft} \times 2 \text{ ft/s}}}$$

$$C_{max} = 103 \text{ ppm}$$

10.3.4 Longitudinal dispersion of an instantaneous point source

Dispersion is caused primarily by the nonuniformity of the velocity profile in the cross section of shear flow. The dispersion coefficient K_d describes the diffusive property of the velocity distribution and is obtained from:

$$K_d = \frac{-1}{h\epsilon} \int_0^h (v_x - V) \int_0^z \int_0^z (v_x - V) \, dz \, dz \, dz \tag{10.13}$$

where v_x is the local velocity, V the depth-averaged flow velocity, h the flow depth, and ϵ the turbulent mixing coefficient. Practical estimates of K_d are given by Equation (10.8d). Washload transport in the streamwise direction is proportional to the concentration gradient and the dispersion coefficient.

Dispersion of a mass of fine sediment added as a point source at a given instant can be effectively calculated after complete vertical mixing, $t > t_v$, and complete transverse mixing, $t > t_t$. The cloud of sediment will approach a Gaussian distribution only after an initial period of dispersion $t_{di} > 0.4W^2/\epsilon_t$. After this initial period, a cloud of total length approximately equal to $4\sigma_d = 4\sqrt{2K_d t}$ will propagate in the downstream direction at a peak concentration given by

$$C_{max} = \frac{m}{Wh\sqrt{4\pi K_d t}} \tag{10.14}$$

Example 10.3 outlines the application of this method to the longitudinal dispersion of an instantaneous point source.

Example 10.3   Application to longitudinal dispersion of a point source.
Localized mass wasting of an overhanging streambank causes a 6,000-kg block of very fine silt to dissolve into a 20-m-wide and 1-m-deep stream. Assuming rapid erosion of the block under a 0.3-m/s flow velocity and a stream slope of 10 cm/km, determine the peak concentration and the

length of the dispersed sediment cloud observed at a bridge located $X_b = 10$ km downstream from the sediment source.

Step 1. The shear velocity is

$$u_* = \sqrt{ghS} = \sqrt{\frac{9.8 \text{ m}}{\text{s}^2} \times 1 \text{ m} \times 10^{-4}} = 0.031 \frac{\text{m}}{\text{s}}$$

Step 2. The dispersion coefficient K_d and the transverse mixing coefficient ϵ_t are, respectively,

$$K_d \cong 250hu_* = 250 \times \frac{1 \text{ m} \times 0.031 \text{ m}}{\text{s}} = \frac{7.8 \text{ m}^2}{\text{s}}$$

$$\epsilon_t = 0.6hu_* = 0.6 \times 1 \text{ m} \times \frac{0.031 \text{ m}}{\text{s}} = \frac{0.0186 \text{ m}^2}{\text{s}}$$

Step 3. The initial period ends when

$$t_{di} = \frac{0.4W^2}{\epsilon_t} = \frac{0.4 \times (20 \text{ m})^2 \text{ s}}{0.0186 \text{ m}^2} = 8{,}602 \text{ s}$$

at a downstream distance X_{di}:

$$X_{di} = Vt_{di} = \frac{0.3 \text{ m}}{\text{s}} \times 8{,}602 \text{ s} = 2{,}580 \text{ m} < 10 \text{ km}$$

Step 4. At the bridge, the length L of the dispersed cloud,

$$L = 4\sqrt{\frac{2K_d X_b}{V}} = 4\sqrt{2 \times \frac{7.8 \text{ m}^2}{\text{s}} \times \frac{10{,}000 \text{ m} \times \text{s}}{0.3 \text{ m}}} = 2.9 \text{ km}$$

will be centered at the bridge at a time $t_b = X_b/V = 10$ km $\times$ s/0.3 m = 9.3 h after the injection upstream.

Step 5. The maximum concentration of the dispersed cloud under the bridge at time t_b is given by

$$C_{max} = \frac{m}{Wh\sqrt{4\pi K_d t_b}} = \frac{6{,}000 \text{ kg}}{20 \text{ m} \times 1 \text{ m} \sqrt{4\pi \times 7.8 \text{ m}^2/\text{s} \times 33{,}333 \text{ s}}}$$

$$= \frac{0.167 \text{ kg}}{\text{m}^3} = 167 \frac{\text{mg}}{\text{l}}$$

10.4 Suspended sediment concentration profile

Consider steady uniform turbulent flow in a wide rectangular channel without any internal sediment source ($\dot{m} = 0$). The deposition of sediment particles due to the density difference between the particles and the surrounding fluid induces a downward particle settling flux. All the terms of Equation (10.5) vanish except those describing the sediment fluxes in the vertical z direction:

$$v_z \frac{\partial C}{\partial z} = \epsilon_z \frac{\partial^2 C}{\partial z^2} \tag{10.15}$$

This equation can be integrated with respect to z. Although the net vertical velocity is zero, the downward fall velocity of sediment particles, $v_z = -\omega$, in Equation (10.15) gives

$$\omega C + \epsilon_z \frac{\partial C}{\partial z} = 0 \tag{10.16}$$

The resulting concentration profile, given constant values of ω and ϵ_z, is simply

$$C = C_0 e^{-\omega z/\epsilon_z} \tag{10.17}$$

This relationship shows that, as $\omega z < \epsilon_z$, the concentration profile becomes gradually uniform, whereas most of the sediment is concentrated near the bed when $\omega z > \epsilon_z$.

For the most general case with variable parameters, the turbulent mixing coefficient of sediment ϵ_z is examined by analogy with the momentum exchange coefficient ϵ_m defined in Equation (6.8):

$$\epsilon_z = \beta_s \epsilon_m = \beta_s \frac{\tau}{\rho_m} \frac{dz}{dv_x} \tag{10.18}$$

in which β_s is the ratio of the turbulent mixing coefficient of sediment to the momentum exchange coefficient. This coefficient remains sufficiently close to unity for practical applications.

The vertical sediment concentration can be determined after substituting the relationships for shear stress [Eq. (8.6)] and for turbulent velocity profiles [Eq. (6.10)] into Equation (10.18) to give

$$\epsilon_z = \beta_s \kappa u_* \frac{z}{h}(h-z) \tag{10.19}$$

The resulting mixing coefficient ϵ_z varies with z. Notice that the maximum value of ϵ_z with $\beta_s = 1$ and $\kappa = 0.4$ equals $\epsilon_z = 0.1 u_* h$ at middepth $z = 0.5h$ [a result quite similar to that of Eq. (10.8a)]. The expression for ϵ_z in Equation (10.19) is substituted into Equation (10.16) and solved after separating the variables C and z:

$$\frac{C}{C_a} = \left(\frac{h-z}{z}\frac{a}{h-a}\right)^{\mathrm{Ro} = \omega/\beta_s\kappa u_*}$$ (10.20) ◆

in which C_a represents the reference sediment concentration at a reference elevation a above the bed elevation. The relative concentration C/C_a depends on the elevation z above the reference elevation. The exponent Ro, referred to as the Rouse number, reflects the ratio of the sediment properties to the hydraulic characteristics of the flow.

With nearly constant values of $\beta_s \simeq 1$ and the von Kármán constant $\kappa \simeq 0.4$, the Rouse number essentially reduces to the ratio of fall velocity to shear velocity. Figure 10.2 illustrates that as the sediment size becomes very small, the sediment concentration profiles become gradually uniform as Ro → 0. Conversely, the concentration of sediment particles becomes increasingly large near the bed as the sediment size increases.

For steady uniform flow, the concentration of suspended sediment C varies with flow depth h and distance above bed z. A log–log plot of C versus $(h-z)/z$ shows a straight line from which the exponent Ro can be graphically defined, as shown in Figure 10.3.

10.5   Suspended load

The unit suspended sediment discharge q_s in natural streams and canals is computed from the depth-integrated advective flux of sediment Cv_x above the bed layer $z > a$:

$$q_s = \int_a^h Cv_x\, dz$$ (10.21)

The corresponding total suspended sediment discharge Q_s is obtained from integration of the unit suspended sediment discharge over the entire width of the channel, or

$$Q_s = \int_{\text{width}} q_s\, dw$$ (10.22)

The suspended load L_s defines the amount of sediment in suspension passing a cross section over a certain period of time; thus,

$$L_s = \int_{\text{time}} Q_s\, dt$$ (10.23)

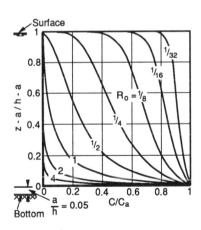

Figure 10.2. Equilibrium sediment concentration profiles for $a/h = 0.05$

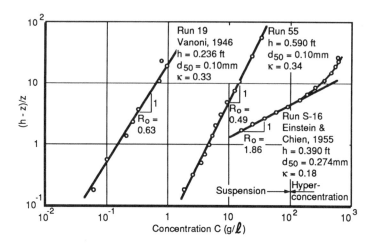

Figure 10.3. Suspended sediment concentration profiles (modified after Woo et al., 1988)

By analogy with bedload, described in Chapter 9, the units can be those of weight, mass, or volume. It is common to use the suspended load in metric tons (1,000 kg).

The comparison of suspended load to bedload delineates which mode of sediment transport is dominant. The suspended unit sediment discharge q_s can be calculated from Equation (10.21) after substituting C from Equation (10.20) and v_x from Equation (6.13):

$$q_s = \int_{z=2d_s}^{h} C_a \frac{u_*}{\kappa} \left[\frac{h-z}{z} \frac{a}{h-a} \right]^{\omega/\beta_s \kappa u_*} \ln \frac{30z}{k_s'} \, dz \qquad (10.24)$$

Similarly, the total unit sediment discharge can be obtained by integrating from $z = k_s'/30$ to the free surface $z = h$; thus,

$$q_t = \int_{z=k_s'/30}^{h} C_a \frac{u_*}{\kappa} \left[\frac{h-z}{z} \frac{a}{h-a} \right]^{\omega/\beta_s \kappa u_*} \ln \frac{30z}{k_s'} \, dz \qquad (10.25)$$

The ratio of suspended to total unit sediment discharges indicates whether most of the sediment transport occurs in suspension or in the bed layer.

After substitution of $\beta_s = 1$, $\kappa = 0.4$, and $k_s' = d_s$, the ratio q_s/q_t from Equations (10.24) and (10.25) becomes independent of C_a and u_*, but varies as a function of the ratio of shear to fall velocities, u_*/ω, and relative submergence h/d_s. Sediment transport can be subdivided into three zones describing the dominant mode of transport: bedload, mixed load, and suspended load. It is interesting that for turbulent flow over rough

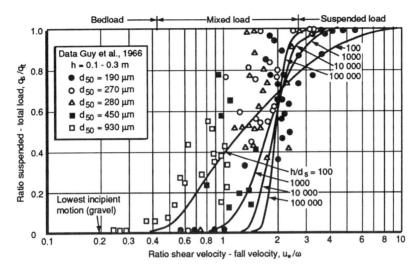

Figure 10.4. Ratio of suspended to total load versus ratio of shear to fall velocities

boundaries, incipient motion corresponds to $u_*/\omega \simeq 0.2$. Figure 10.4 shows the ratio of suspended to total load as a function of u_*/ω and h/d_s. Bedload is dominant at values of u_*/ω less than about 0.4, and the methods detailed in Chapter 9 should be used to determine the sediment transport rate. A transition zone called mixed load is found where $0.4 < u_*/\omega < 2.5$ in which both the bedload and the suspended load contribute to the total load. Methods for calculating the total load are presented in Chapter 11. Suspended load is dominant when $u_*/\omega > 2.5$, and gravitational effects on particles are negligible compared with turbulent mixing. As u_*/ω approaches 2.5, the concentration $C_{0.8}$ at $y = 0.8h$ is about 75% of the concentration $C_{0.2}$ at $y = 0.2h$ ($C_{0.8} = 0.75C_{0.2}$), $C_{0.8} = 0.93C_{0.2}$ when $u_*/\omega = 100$, and $C_{0.8} = 0.98C_{0.2}$ when $u_*/\omega = 400$. The advection–diffusion equations presented in Sections 10.2 and 10.3 are most suitable when $u_* > 100\omega$. This usually corresponds to most silts and all clays. These observations are summarized in Table 10.2.

10.6   Hyperconcentrations

Hyperconcentrations refer to highly sediment-laden flows in which the presence of fine sediments materially affects fluid properties and bed material transport. In general, the volumetric sediment concentration C_v of hyperconcentrations ranges from 5% to 60% according to various measurements. The mass density of hyperconcentrations ρ_m is calculated from

Table 10.2. *Modes of sediment transport given shear and fall velocities*

u_*/ω	Mode of sediment transport
<0.2	No motion for all possible grain sizes
0.2	Lowest possible incipient motion for turbulent flow over rough boundaries
0.2–0.4	Sediment transport as bedload (Chapter 9)
0.4–2.5	Sediment transport as mixed load (Chapter 11)
>2.5	Sediment transport in suspension (Chapter 10)
25	$C_{0.8} = 0.75C_{0.2}$
100	$C_{0.8} = 0.93C_{0.2}$
400	$C_{0.8} = 0.98C_{0.2}$

$\rho_m = \rho + (\rho_s - \rho)C_v$. With a decrease in water content, hyperconcentrations grade into earth flows and dry landslides. Mudflows, debris flows, and hyperconcentrated flows are recognized. A mudflow is a flowing mass of predominantly fine-grained material that possesses a high degree of fluidity during movement. As a rule of thumb, if more than half of the solid fraction is coarser than sand, the term *debris flow* is preferable. Debris flows are characterized by poorly sorted deposits supporting coarse clasts in a fine matrix, while hyperconcentrated flows do not necessarily have the percentage of clay needed to exhibit yield strength. It is generally accepted that mudflows contain sufficient proportions of fines, silts, and clays to alter the fluid properties in terms of viscosity and yield strength.

10.6.1   Rheology of hyperconcentrations

Rheology is a science describing the deformation and flow of matter. More specifically, the graphical measure of the shear stress applied at a given rate of deformation of a fluid defines a rheogram (Fig. 10.5). In clear-water flows, the shear stress increases linearly with the rate of deformation in the laminar flow regime and the fluid is said to be Newtonian [e.g., Eqs. (2.2) and (5.1)]. The dynamic viscosity of a mixture μ_m is then defined as the slope of the rheogram. Under large shear stresses, the boundary layer flow becomes turbulent (except in the laminar sublayer), and the shear stress increases with the second power of the rate of deformation.

The Bingham rheological model is to some extent a limiting or idealized rheological model. Beyond a finite shear stress, called *yield stress* τ_y, the rate of deformation, dv_x/dz, is linearly proportional to the excess shear stress. The constitutive equation is

$$\tau = \tau_y + \mu_m \frac{dv_x}{dz} \tag{10.26}$$

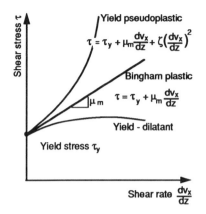

Figure 10.5. Rheogram for non-Newtonian fluids

in which μ_m is the dynamic viscosity of the mixture. The Bingham plastic model is well suited to homogeneous suspensions of fine particles, particularly at low rates of deformation. Experimental laboratory results by Qian and Wan (1986) and others confirm that under rates of deformation observed in the field, fluids with large concentrations of fine particles behave like Bingham plastic fluids.

The analysis of coarse sediment mixtures is somewhat more complex and involves an additional shear stress due to particle impact. Bagnold (1954) pioneered laboratory investigations on the impact of sediment particles. He defined the dispersive shear stress τ_d induced by the collision between sediment particles as

$$\tau_d = c_{Bd}\rho_s\left[\left(\frac{0.615}{C_v}\right)^{1/3} - 1\right]^{-2} d_s^2\left(\frac{dv_x}{dz}\right)^2 \tag{10.27}$$

The dispersive shear stress is shown to increase with three parameters: the second power of the particle size, the volumetric sediment concentration, and the second power of the rate of deformation. It is important to recognize that the dispersive stress is proportional to the product of these three parameters; therefore, high values of all parameters are required to induce a significant dispersive shear stress.

A quadratic rheological model has been proposed by O'Brien and Julien (1985) that combines the following stress components of hyperconcentrated sediment mixtures: (1) cohesion between particles, (2) internal friction between fluid and sediment particles, (3) turbulence, and (4) inertial impact between particles. The resulting quadratic model is

$$\tau = \tau_y + \mu_m\frac{dv_x}{dz} + \zeta\left(\frac{dv_x}{dz}\right)^2 \tag{10.28}$$

where μ_m is the dynamic viscosity of the mixture and ζ is the turbulent-dispersive parameter. The last term of the quadratic model combines the effects of turbulence with the dispersive stress induced by the inertial impact of sediment particles. Combining the conventional expression for the turbulent stress in sediment-laden flows with Bagnold's dispersive stress gives

$$\zeta = \rho_m l_m^2 + c_{Bd}\rho_s \left[\left(\frac{0.615}{C_v} \right)^{1/3} - 1 \right]^{-2} d_s^2 \tag{10.29}$$

where ρ_m and l_m are, respectively, the mass density and mixing length of the mixture, d_s is the particle diameter, and c_{Bd} is an empirical parameter defined by Bagnold ($c_{Bd} \cong 0.01$).

10.6.2   Parameter evaluation

The rheological properties of hyperconcentrations are determined from laboratory analyses of rheograms obtained from viscometric measurements. At least three kinds of device are commercially available for the measurement of rheograms: the capillary viscometer, the concentric cylindrical viscometer, and the cone and plate viscometer. Concentric cylindrical viscometers seem to be best suited for a wide range of shear rates. Field observations, however, indicate that shear rates rarely exceed 100/s. It is therefore recommended that the rheological properties of hyperconcentrations be measured under low rates of shear. Rheological properties of hyperconcentrations are generally formulated as a function of sediment concentration, as shown in Figure 10.6. The recommended empirical formulas are the exponential relationships for yield stress and viscosity at large concentrations of fines, as well as the Bagnold equation to calculate the dispersive shear stress of coarse particles.

The yield strength τ_y of a mixture ($C_v > 0.05$) can be approximated by

$$\tau_y \text{ (in Pa)} \cong 0.1e^{3(C_v - 0.05)} \quad \text{for sands} \tag{10.30a}$$

$$\tau_y \text{ (in Pa)} \cong 0.1e^{13(C_v - 0.05)} \quad \text{for 95\% silts and 5\% clays} \tag{10.30b}$$

$$\tau_y \text{ (in Pa)} \cong 0.1e^{23(C_v - 0.05)} \quad \text{for 70\% silts and 30\% clays} \tag{10.30c}$$

The corresponding dynamic viscosity of the mixture μ_m is given by

$$\mu_m \cong \mu(1 + 2.5C_v + e^{10(C_v - 0.05)}) \quad \text{for sands} \tag{10.31a}$$

$$\mu_m \cong \mu(1 + 2.5C_v + e^{23(C_v - 0.05)}) \quad \text{for silts and clays} \tag{10.31b}$$

Based on limited laboratory data, the turbulent-dispersive parameter ζ is negligible for silts and clays, and for sands it can be approximated by

$$\zeta \text{ (in kg/m)} \cong 10^{-6}e^{15(C_v - 0.05)} \quad \text{for sands} \tag{10.32}$$

10.6.3   Fall velocity and particle buoyancy

The fall velocity of particles of diameter d_s and specific weight γ_s in a Bingham plastic fluid of specific weight γ_m is given by

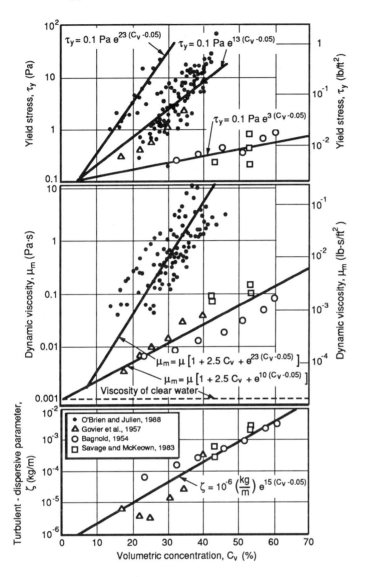

Figure 10.6. Physical properties of hyperconcentrations (modified after Julien and Lan, 1991)

$$\omega^2 = \frac{4}{3} \frac{gd_s}{C_D} \frac{\gamma_s - \gamma_m}{\gamma_m} \tag{10.33}$$

where the drag coefficient C_D depends on the dynamic viscosity μ_m and the yield stress τ_y of the Bingham fluid. After defining the Bingham Reynolds number $\mathrm{Re_B} = \rho_m d_s \omega / \mu_m$ and the Hedstrom number $\mathrm{He} = \rho_m d_s^2 \tau_y / \mu_m^2$, one can rewrite the drag coefficient as follows:

$$C_D = \frac{24}{\mathrm{Re_B}} + \frac{2\pi \mathrm{He}}{\mathrm{Re_B^2}} + 1.5 \tag{10.34}$$

The settling velocity ω can be obtained from Equations (10.33) and (10.34) as a function of particle diameter d_s, yield strength τ_y, mixture viscosity μ_m, and mass density ρ_m:

$$\omega = \frac{8\mu_m}{\rho_m d_s}\left[\left(1 + \frac{\rho_m g d_s^3 (\rho_s - \rho_m)}{72\mu_m^2} - \frac{\pi}{48}\frac{\rho_m \tau_y d_s^2}{\mu_m^2}\right)^{0.5} - 1\right] \tag{10.35}$$

This relationship reduces to Equation (5.23c) when $\tau_y = 0$, $\rho_m = \rho$, and $\mu_m = \mu$. Interestingly, the fall velocity reduces to zero for particles smaller than the buoyant particle diameter d_{sb} given by

$$d_{sb} = \frac{3\pi}{2}\frac{\tau_y}{\gamma_s - \gamma_m} \tag{10.36}$$

The fine particles $d_s < d_{sb}$ remain neutrally buoyant in the mixture and do not settle.

10.6.4 Dimensionless rheological model

The relative magnitude of the terms in the quadratic equation [Eq. (10.28)] define conditions under which simplified rheological models are applicable. The dimensionless rheological model of Julien and Lan (1991) is obtained after rewriting Equations (10.28) and (10.29) as

$$\Pi_\tau = 1 + (1 + \Pi_{td}) c_{Bd}\Pi_{dv} \tag{10.37}$$

in which the dimensionless parameters are defined as

1. Dimensionless excess shear stress Π_τ:

$$\Pi_\tau = \frac{\tau - \tau_y}{\mu_m \, dv_x / dz} \tag{10.38a}$$

2. Dimensionless dispersive–viscous ratio Π_{dv}:

$$\Pi_{dv} = \frac{\rho_s d_s^2}{\mu_m}\left[\left(\frac{0.615}{C_v}\right)^{1/3} - 1\right]^{-2}\left(\frac{dv_x}{dz}\right) \tag{10.38b}$$

3. Dimensionless turbulent–dispersive ratio Π_{td}:

$$\Pi_{td} = \frac{\rho_m l_m^2}{c_{Bd}\rho_s d_s^2}\left[\left(\frac{0.615}{C_v}\right)^{1/3} - 1\right]^2 \tag{10.38c}$$

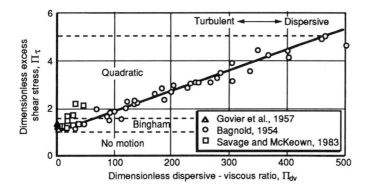

Figure 10.7. Classification of rheological models
(modified after Julien and Lan, 1991)

The usefulness of the dimensionless rheological model is demonstrated in Figure 10.7, where Π_τ is plotted versus Π_{dv}. It is interesting that when Equation (10.37) is fitted to the experimental data sets of Govier et al. (1957), Savage and McKeown (1983), and Bagnold (1954), the value $c_{Bd} = 0.01$ suggested by Bagnold is comparable to the value $c_{Bd}(1 + \Pi_{rd}) = 0.0087$ obtained from the slope of the line in Figure 10.7.

Simplifications of the quadratic rheological model are possible under the following conditions: (1) The Bingham model [Eq. (10.26)] is applicable when $\Pi_\tau \rightarrow 1$; moreover, the fluid is Newtonian when $\tau_y \ll \tau$. (2) The flow is turbulent when $\Pi_\tau > 5$ and $\Pi_{td} > 1$. (3) The dispersive shear stress is dominant when $\Pi_\tau > 5$ and $\Pi_{dv} > 400$.

This analysis suggests that hyperconcentrations could be classified as (1) mudflows when the Bingham rheological model is applicable; (2) hyperconcentrated flows when the turbulent shear stress is dominant; or (3) debris flows when the dispersive stress predominates. Example 10.4 provides detailed calculations of some hyperconcentrated flow characteristics. Example 10.5 provides a similar example for mudflows.

Example 10.4   Application to hyperconcentrated flow. A mixture of uniform medium sand, $d_{50} = 0.5$ mm, flows on a very steep slope, $S_0 = 0.25$. If the flow depth is 40 cm and the volumetric concentration is $C_v = 0.4$, estimate the following: (1) the mass density of the mixture; (2) the yield stress; (3) the kinematic viscosity of the mixture; (4) the turbulent-dispersive parameter; (5) the buoyant particle diameter; (6) the fall velocity of 0.1- and 1-mm sand; and (7) which rheological model applies when $dv_x/dz = 50/s$.

Step 1.

$$\rho_m = \rho(1+(G-1)C_v) = \frac{1,000\ \text{kg}}{\text{m}^3}(1+(2.65-1)0.4) = \frac{1,660\ \text{kg}}{\text{m}^3}$$

Step 2. From Equation (10.30a):

$$\tau_y \cong 0.1e^{3(0.4-0.05)} = 0.28\ \text{Pa}$$

Step 3. From Equation (10.31a) and μ at $20°\text{C}$,

$$\mu_m \cong 1\times10^{-3}[1+(2.5\times0.4)+e^{10(0.04-0.05)}] = 0.035\ \frac{\text{N}\times\text{s}}{\text{m}^2}$$

Step 4. From Equation (10.32),

$$\zeta \cong 10^{-6}e^{15(0.4-0.05)} = 1.9\times10^{-4}\ \text{kg/m}$$

Step 5. From Equation (10.36),

$$d_{sb} = \frac{3\pi\tau_y}{2(\rho_s-\rho_m)g} = \frac{3\pi 0.28\ \text{N}\times\text{m}^3\times\text{s}^2}{2\ \text{m}^2(2,650-1,660)\ \text{kg}\ 9.81\ \text{m}} = 0.14\ \text{mm}$$

Step 6. (a) $\omega = 0$ from 0.1-mm sand because $d_s < d_{sb} = 0.14$ mm; (b) the fall velocity from Equation (10.35) for 1-mm sand is

$$\omega = \frac{8\mu_m}{\rho_m d_s}\left[\left(1+\frac{\rho_m g d_s^3}{72\mu_m^2}(\rho_s-\rho_m)-\frac{\pi}{48}\frac{\rho_m\tau_y d_s^2}{\mu_m^2}\right)^{0.5}-1\right]$$

$$= \frac{8\times0.035\ \text{N}\times\text{s}\times\text{m}^3}{\text{m}^2\times1,660\ \text{kg}\ 0.001\ \text{m}}$$

$$\times\left[\left(1+\frac{1,660\ \text{kg}\times9.81\ \text{m}(0.001)^3\ \text{m}^3\times\text{m}^4(2,650-1,660)\ \text{kg}}{72\ \text{m}^3\times\text{s}^2(0.035)^2\ \text{N}^2\times\text{s}^2\times\text{m}^3}\right.\right.$$

$$\left.\left.-\frac{\pi}{48}\frac{1,660\ \text{kg}}{\text{m}^3}\frac{0.28\ \text{N}}{\text{m}^2}\frac{(0.001)^2\ \text{m}^2\times\text{m}^4}{(0.035)^2\ \text{N}^2\times\text{s}^2}\right)^{0.5}-1\right] = \frac{0.0128\ \text{m}}{\text{s}}$$

Step 7. The applied boundary shear τ_0 at $dv_x/dz = 50/\text{s}$ is

$$\tau_0 = \rho_m g h S_0 = \frac{1,660\ \text{kg}}{\text{m}^3}\times\frac{9.8\ \text{m}}{\text{s}^2}\times0.4\ \text{m}\times0.25 = 1,628\ \text{Pa}$$

The dimensionless parameters from Equation (10.38) give

$$\Pi_\tau = \frac{\tau_0-\tau_y}{\mu_m\ dv_x/dz} = \frac{(1,628-0.28)\ \text{N}\times\text{m}^2\times\text{s}}{\text{m}^2\times0.035\ \text{N}\times\text{s}\times50} = 930$$

Both the viscous and yield stresses are negligible.

Assuming $l_m = 0.4h$ and $c_{Bd} = 0.01$,

$$\Pi_{td} = \frac{\rho_m(0.4)^2 h^2}{c_{Bd}\rho_s d_{50}^2}\left[\left(\frac{0.615}{C_v}\right)^{1/3}-1\right]^2$$

$$= \frac{1,660 \text{ kg}}{\text{m}^3} \frac{(0.4)^2 (0.4)^2 \text{ m}^2 \times \text{m}^3}{0.01 \times 2,650 \text{ kg} (0.0005)^2 \text{ m}^2} \left[\left(\frac{0.615}{0.4} \right)^{1/3} - 1 \right]^2 = 152,472$$

The dispersive stress is negligible. A turbulent model should be applicable.

Example 10.5 Application to mudflow. A 0.4-m-thick layer of mud flows down a steep mountain channel, $S_0 = 0.05$. The analysis of a sample reveals that the total volumetric concentration $C_v = 0.5$, and the sample is composed of 5% clay, 75% silt, and 20% sand with $d_{50} = 0.05$ mm. Estimate the following: (1) mass density of the mixture; (2) volumetric concentration of fines; (3) yield stress; (4) kinematic viscosity of the mixture; (5) buoyant particle diameter; (6) fall velocity of 1.0-mm sand and 10-mm gravel; and (7) which rheological model is likely to apply when $dv_x/dz = 50/\text{s}$.

Step 1.
$$\rho_m = \rho(1 + (G-1)C_v) = \frac{1,000 \text{ kg}}{\text{m}^3}(1 + (2.65 - 1)0.5) = \frac{1,825 \text{ kg}}{\text{m}^3}$$

Step 2. The concentration C_{vM} of silts and clays in the fluid matrix,
$$C_{vM} = C_v(1 - \% \text{ sand}) = 0.5(1 - 0.2) = 0.4$$

Step 3. The silt–clay ratio suggests the use of Equation (10.30b):
$$\tau_y \cong 0.1e^{13(C_{vM} - 0.05)} = 9.5 \text{ Pa}$$

Step 4. From Equation (10.31b) and μ at 20°C,
$$\mu_m \cong \mu(1 + 2.5C_{vM} + e^{23(C_{vM} - 0.05)})$$
$$\cong \frac{1 \times 10^{-3} \text{ N} \times \text{s}}{\text{m}^2}(1 + (2.5 \times 0.4) + e^{23(0.4 - 0.05)})$$
$$\mu_m \cong \frac{3.136 \text{ N} \times \text{s}}{\text{m}^2}$$

Step 5.
$$d_{sb} = \frac{3\pi}{2} \frac{\tau_y}{(\rho_s - \rho_m)g} = \frac{3\pi \times 9.5 \text{ Pa} \times \text{m}^3 \text{s}^2}{2(2,650 - 1,825) \text{ kg} \times 9.81 \text{ m}} = 5.5 \text{ mm}$$

Step 6. $\omega = 0$ for 1-mm sand because $d_s < d_{sb} = 5.5$ mm. From Equation (10.35), the settling velocity for 10-mm gravel is
$$\omega = \frac{8\mu_m}{\rho_m d_s} \left[\left(1 + \frac{\rho_m g d_s^3 (\rho_s - \rho_m)}{72\mu_m^2} - \frac{\pi}{48} \frac{\rho_m \tau_y d_s^2}{\mu_m^2} \right)^{0.5} - 1 \right]$$
$$\omega = \frac{8 \times 3.136 \text{ N} \times \text{s} \times \text{m}^3}{\text{m}^2 \times 1,825 \text{ kg } 0.01 \text{ m}}$$

$$\times \left[\left(1 + \frac{1{,}825 \text{ kg}}{\text{m}^3} \frac{9.81 \text{ m}}{\text{s}^2 \times \text{m}^3} \frac{(0.01)^3 \text{ m}^3 \, (2{,}650 - 1{,}825) \text{ kg} \times \text{m}^4}{72(3.136)^2 \text{ N}^2 \times \text{s}^2} \right. \right.$$

$$\left. \left. - \frac{\pi}{48} \frac{1{,}825 \text{ kg}}{\text{m}^3} \frac{9.5 \text{ N}}{\text{m}^2} \frac{(0.01)^2 \text{ m}^2 \times \text{m}^4}{(3.136)^2 \text{ N}^2 \times \text{s}^2} \right)^{0.5} - 1 \right]$$

$$= \frac{1.37 \text{ m}}{\text{s}} [(1 + .0208 - .0115)^{0.5} - 1] = 6.4 \frac{\text{mm}}{\text{s}}$$

Notice that the hyperconcentrated sediment mixture reduced the settling velocity of a 10-mm diameter gravel particle to an equivalent fine sand particle in clear water.

Step 7. The applied boundary shear τ_0 is

$$\tau_0 = \rho_m g h S_0 = \frac{1{,}825 \text{ kg}}{\text{m}^3} \times \frac{9.81 \text{ m}}{\text{s}^2} \times 0.4 \text{ m} \times 0.05 = 358 \text{ Pa}$$

The dimensionless parameters from Equation (10.38) give

$$\Pi_\tau = \frac{\tau - \tau_y}{\mu_m \, dv_x/dz} = \frac{(358 - 9.5) \text{ Pa}}{3.136 \text{ N} \times \text{s} \times 50} \text{ m}^2 \times \text{s} = 2.2$$

The yield stress is small and the viscous stress is significant:

$$\Pi_{dv} = \frac{\rho_s d_{50}^2}{\mu_m} \left[\left(\frac{0.615}{C_v} \right)^{1/3} - 1 \right]^{-2} \frac{dv_x}{dz}$$

$$= \frac{2{,}650 \text{ kg}}{\text{m}^3} \frac{(5 \times 10^{-5})^2 \text{ m}^2}{3.136 \text{ N} \times \text{s}} \text{ m}^2 \left[\left(\frac{0.615}{0.5} \right)^{1/3} - 1 \right]^{-2} \frac{50}{\text{s}} = 0.02$$

The dispersive stress is very small. Turbulence may be significant. A quadratic model from Equation (10.28) is therefore indicated.

10.7   Field measurements of suspended sediment

The quantity of sediments held in suspension in a stream can be measured from representative samples of the water–sediment mixture. Three types of samplers are considered.

10.7.1   *Instantaneous samplers*

Instantaneous samplers trap a volume of the suspension flowing through a cylindrical tube by simultaneously closing off both ends. The sample is then filtered and dried to provide a measure of volume-averaged sediment concentration C_v.

Instruments of a container type that can be opened and closed instantaneously have been designed. Studies of their effectiveness indicated that they are not suitable for general field use. They are best used in reservoir studies and at river locations where the intensity of turbulence is sufficiently high to maintain all the sediment particles in suspension.

10.7.2 Point samplers

Point samplers are designed to collect through time a sample at a given point in the stream vertical. The dried sample measures the time-averaged sediment concentration C_t.

The body of the sampler contains air, which is compressed by the inflowing liquid so that its pressure balances the external hydrostatic head. A rotary valve that opens and closes the sampler is remotely operated. During the sampling period, the valve is opened and the air escapes from the sampler at a nozzle intake velocity nearly equal to the local stream velocity.

Point samplers such as the P-61 (100 lb), P-63 (200 lb), and P-50 (300 lb) are commonly used to obtain information on sediment distribution along a vertical. The P-61 shown in Figure 10.8a uses a pint milk bottle container, while the P-63 (Fig. 10.8b) is designed to use either a pint or a quart sample container. The P-50 is very similar to the P-63 and has been developed for use on major streams such as the lower Mississippi River. The capacity of the container is a quart.

10.7.3 Depth-integrating samplers

Integrating samplers move vertically at a constant speed with an upstream-pointed nozzle. The sampling intake velocity equals the natural flow velocity at every point. The sample is then dried to measure to flux-averaged concentration C_f.

Depth-integrating samplers accumulate water and sediment as they are lowered or raised along the vertical at a uniform rate. The air in the sample container is compressed by the inflowing liquid. If the lowering speed is such that the rate of air compression exceeds the normal rate of liquid inflow, the actual rate of inflow will exceed the local stream velocity and inflow may occur through the air exhaust. If the transit speed is too low, the container will be filled before the total depth is reached. In practice, the transit time must be adjusted so that the container is not completely filled. The descending and ascending velocities need not be equal, but the velocity must remain constant during each phase.

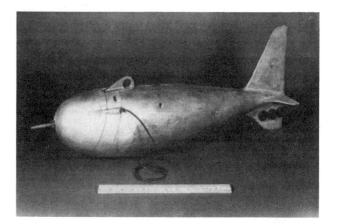

(a)

(b)

Figure 10.8. Point-integrating samplers: (a) P-61; (b) P-63
(from Edwards and Glysson, 1988)

Specifically, the lowering rate should not exceed 40% of the mean flow velocity to avoid excessive angles between the nozzle and the approaching flow. Laboratory tests provide useful information on the recommended lowering rates R_T as a function of the flow velocity and the flow depth for a 1-pint container in a depth-integrating sampler with three intake nozzle diameters (Fig. 10.9).

The DH-48 sampler (4.5 lb) in Figure 10.10a is used in shallow flows when the unit discharge does not exceed 1 m²/s. The DH-59 (24 lb; Fig. 10.10b) is used in streams with low velocities and depths beyond the wading range. The D-49 (62 lb) with cable suspension is designed for use in streams beyond the range of hand-operated equipment.

10.7.4 Sediment discharge measurements

Sediment transport in natural streams is very important, and the accuracy of sediment discharge measurements depends not only on the field methods and equipment utilized for data collection, but also on representative measurements of the sediment distribution in the flow. In natural streams, the sediment concentration at a given cross section varies in both the vertical and the transversal directions and also changes with time. Therefore, the concentration of sediment in suspension may increase or decrease in both space and time, although the flow remains steady in a straight uniform reach.

The standard procedure for measuring the rate of sediment transport at a given instant is essentially based on the definition of the advective flux-averaged concentration [Eq. (10.12c)]; hence,

$$Q_s = C_f Q = \int_A C v_x \, dA \qquad (10.39)$$

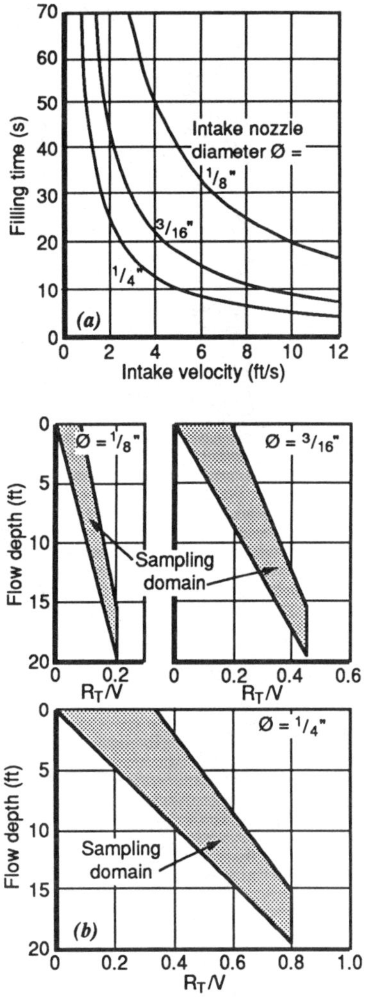

Figure 10.9. Depth-integrating sampling characteristics

in which A is the cross-sectional area, C_f the flux-averaged concentration, and v_x the velocity of the sediment particles.

In small streams it is sometimes possible to measure the total stream discharge by using (1) volumetric methods, (2) the dilution method, or (3) the weir equation. In the first method, if the total flow of a small stream can be directed into a given basin or container of fixed volume, the ratio of the volume to the filling time determines the time-averaged discharge entering the container.

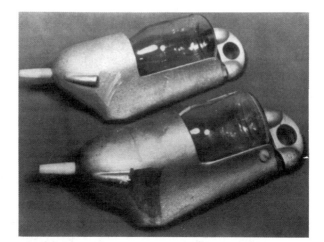

(a)

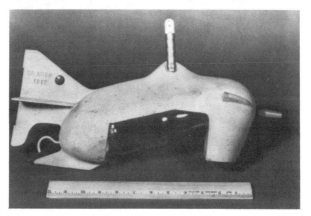

(b)

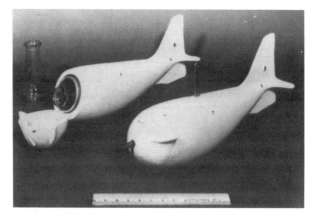

(c)

Figure 10.10. Depth-integrating samplers: (a) DH-48; (b) DH-59;
(c) D-74 (from Edwards and Glysson, 1988)

With the second method, a dilute suspension at concentration C_1 of a conservative substance is steadily injected at a constant rate Q_1 into a small turbulent stream of unknown discharge. Note that C_1 represents the difference between the concentration of the solution and the natural concentration of the stream. With reference to Section 10.3, a sample of water is taken at a certain distance downstream from the source point where the vertical and transversal mixing is completed. With a measure of the uniform concentration C_2 of the sample, the discharge Q_2 of the stream can be obtained simply by $Q_2 = C_1 Q_1 / C_2$.

With the third method, man-made structures with a fixed geometry, such as weirs, spillways, and pipes, are first located. The discharge through these elements is then evaluated from the geometry, stage, and velocity measurements. For example, the weir equation can be used with simple width and stage measurements.

The flux-averaged concentration in small streams can often be measured at locations where the turbulence intensity is very high, such as in plunge pools, downstream from contractions, or in turbulence flumes. Particles from fractions of all sizes are held in suspension, and a bulk sample provides an accurate evaluation of the flux-averaged sediment concentration. Turbulence flumes consisting of a series of baffles anchored to a concrete slab can be installed. The turbulence induced by the baffles is sufficient to transport in suspension almost the entire load in the stream. In streams where mostly fine sediments (silts and clays) are held in suspension, the exponent of the Rouse equation is generally very small and uniform concentration of sediment can be assumed for those fractions. In such cases, a sample taken at any location along the vertical is sufficient to describe the flux-averaged concentration for those size fractions.

Exercises

*10.1. Derive concentration by weight C_w [Eq. (10.1b)] from concentration by volume C_v [Eq. (10.1a)] given the density of sediment particles $G = \gamma_s / \gamma$.

*10.2. (a) Derive Equation (10.19) from Equation (10.18); and
(b) evaluate the maximum value of ϵ_z on a vertical.

Problem 10.1

Plot the dimensionless concentration profiles C/C_a for medium silt, fine sand, and coarse sand in a 3-m-deep stream sloping at $S_0 = 0.002$. (*Hint:* Assume $a = 2d_s$.)

Answer

$C/C_a > 0.95$ for medium silt;
$C/C_a = 0.36$ at middepth for fine sand;
$C/C_a < 0.05$ for most of the coarse sand profile

Problem 10.2

Given the sediment concentration profile from Problem 6.1:

(a) plot the concentration profile $\log C$ versus $\log(h-z)/z$;
(b) estimate the particle diameter from the Rouse number in (a); and
(c) determine the unit sediment discharge from the given data.

Problem 10.3

Calculate the daily sediment load in a nearly rectangular 50-m-wide stream with an average flow depth $h = 2$ m and slope $S_0 = 0.0002$ when 25% of the sediment load is fine silt, 25% is very fine silt, and 50% is clay, and the middepth concentration is $C = 50,000$ mg/l.

Answer: $L_s \cong 750,000$ tons/d.

*Problem 10.4

A physical model of a river 50 m wide and 2 m deep is to be constructed in the hydraulics laboratory at a scale of 1:100 horizontal and 1:20 vertical. Calculate the ratio of transversal to vertical mixing time scales: (a) for the model and (b) for the prototype.

*Problem 10.5

Calculate the distance required for complete transversal mixing in a large river at a discharge of 500,000 ft^3/s. Assume an average river width of 2 mi, a slope of about 0.4 ft/mi, and Manning coefficient $n = 0.02$.

Answer: More than 30,000 mi!

**Problem 10.6

The tabulation on page 203 gives the velocity distribution and the suspended sand concentration for the fraction passing a 0.105-mm sieve and retained on a 0.074-mm sieve on the Missouri River. Given the slope 0.00012, flow depth 7.8 ft, river width 800 ft, and water temperature 7°C:

(a) plot the velocity profile V versus $\log z$ and concentration $\log C$ versus $\log(h - z)/z$;

(b) compute from the graphs and given data the following:

$u_* =$ shear velocity,
$V =$ mean velocity,
$\kappa =$ von Kármán constant,
$f =$ Darcy–Weisbach friction factor, and
Ro = Rouse number;

(c) compute the unit sediment discharge for this size fraction from field measurements; and

(d) calculate the flux-averaged concentration.

Distance from bed (ft)	Velocity (ft/s)	Concentration (mg/l)
0.7	4.3	411
0.9	4.5	380
1.2	4.64	305
1.4	4.77	299
1.7	4.83	277
2.2	5.12	238
2.7	5.30	217
2.9	5.40	—
3.2	5.42	196
3.4	5.42	—
3.7	5.50	184
4.2	5.60	—
4.8	5.60	148
5.8	5.70	130
6.8	5.95	—

*Problem 10.7

Bank erosion of fine silts occurs on a short reach of a 100-m-wide meandering river at a discharge of 750 m³/s. If the riverbed slope is 50 cm/km and the flow depth is 5 m, and if the mass wasted is of the order of 10 metric tons per hour, determine the distance required for complete mixing in the river, the maximum concentration at that point, and the average sediment concentration.

Answer: $X_t = 1,875$ m; $C_{max} = 4.3$ mg/l; $C_{av} = 3.7$ mg/l.

11

Total load

Every sediment particle that passes a given stream cross section must satisfy the following two conditions: (1) It must have been eroded somewhere in the watershed above the cross section; and (2) it must be transported by the flow from the place of erosion to the cross section. To this statement, Einstein (1964) added that each of these two conditions may limit the rate of sediment transport depending on the relative magnitude of two controls: (1) the transport capacity of the stream and (2) the availability of material in the watershed. The amount of material transported in a stream therefore depends on two groups of variables: (1) those governing the sediment transport capacity of the stream such as channel geometry, width, depth, shape, wetted perimeter, alignment, slope, vegetation, roughness, velocity distribution, tractive force, turbulence, and uniformity of discharge; and (2) those reflecting the quality and quantity of material made available for transport by the stream, including watershed topography, geology, the magnitude, intensity, and duration of rainfall and snowmelt, weathering, vegetation, cultivation, grazing and land use, soil type, particle size, shape, specific gravity, resistance to wear, settling velocity, mineralogy, cohesion, surface erosion, bank cutting, and sediment supply from tributaries.

A quantitative analysis of the amount of sediment supplied to a stream from a watershed is usually difficult to perform because of the complexity of the physical processes involved and the spatial and temporal variability of all the parameters describing upland erosion from local rainstorms and bank erosion processes. Sediment transport capacity, however, is amenable to quantified analysis and interpretation. The sediment transport capacity of a stream under unlimited sediment supply can be determined as a function of the hydraulic variables and the shape of the stream cross section. The total sediment load in a stream can be divided in three ways:

1. *By the type of movement.* The total sediment load L_T can be divided into the bedload L_b (treated in Chapter 9) and the suspended sediment load L_s (covered in Chapter 10):

$$L_T = L_b + L_s \tag{11.1a}$$

2. *By the method of measurement.* In this case, the total sediment load L_T consists of the measured sediment load L_m and the unmeasured sediment load L_u. Because one can sample from the water surface to a distance of only approximately 10 cm above the bed surface, the measured sediment load L_m is only part of the suspended load L_s. The unmeasured sediment load L_u consists of the entire bedload L_b plus the fraction of the suspended load L_s transported below the lowest sampling elevation:

$$L_T = L_m + L_u \tag{11.1b}$$

3. *By the source of sediment.* In this case (Fig. 11.1), the total sediment load L_T is equal to the supply-limited washload L_w plus the capacity-limited bed-material load L_{bm}. In general, the washload covers the range of fine particles not found in large quantities in the bed ($d_s < d_{10}$), for which sediment transport is limited by the upstream supply of fine particles. The bed-material load is determined entirely by the capacity of the flow to transport the sediment sizes found in the bed:

$$L_T = L_w + L_{bm} \tag{11.1c}$$

It must be recognized that it is virtually impossible to determine the total sediment load of a stream from any of Equations (11.1).

In Section 11.1, several sediment transport formulas commonly used in engineering practice are described with detailed calculation examples. Section 11.2 focuses on supply-limited sediment transport in which sediment sources and sediment yield from upland areas provide the main source of sediments. Section 11.3 describes sediment-rating curves; Section 11.4 covers short- and long-term sediment load. This chapter includes three case studies of capacity- and supply-limited sediment transport.

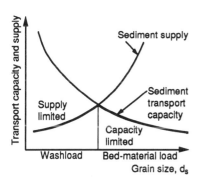

Figure 11.1. Sediment transport capacity and supply curves

11.1   Sediment transport capacity

Numerous sediment transport formulas have been proposed in the past fifty years, and subsequent modifications of original formulations have been prescribed. Although significant progress has been made, none of the existing sediment transport formulas truly fulfills its task. In engineering practice, one compares several formulas with field observations to select the most appropriate equation at a given field site.

For given streamflow conditions, a sediment transport equation can only *predict* the sediment transport capacity of a given bed sediment mixture. Existing sediment transport formulas can be classified into several categories according to their basic approaches: (1) formulations based on advection–diffusion; (2) formulations based on energy concepts in which the rate of work done for transporting sediment particles in turbulent flow is related to the rate of energy expenditure; and (3) empirical equations based on regression analysis and graphical methods.

The following ten methods reflect past and recent developments in sediment transport calculations. Einstein's method is still viewed as a landmark, despite the complexity of the diffusion-based procedure and some uncertainties regarding arbitrarily determined coefficients. The methods of Colby and of Simons, Li, and Fullerton were derived from Einstein's method and involve simple calculations without significant loss of accuracy. Four methods based on energy and stream power concepts are presented. The methods of Bagnold as well as of Engelund and Hansen offer simplicity, while those of Ackers and White as well as of Yang have gained popularity in computer models. Finally, the methods of Shen and Hung, Brownlie, and Karim and Kennedy are essentially the result of regression analysis using comprehensive data sets.

11.1.1   Einstein's approach

The total bed sediment discharge per unit width q_t can be calculated from the sum of the unit bed sediment discharge q_b and the unit suspended sediment discharge q_s from Equation (10.21). Thus,

$$q_t = q_b + \int_a^h C v_x \, dz \tag{11.2}$$

Given the velocity profile for a hydraulically rough boundary from Equation (6.13) with $k'_s = d_s$ and the Rouse concentration profiles from Equation (10.20), the total unit bed sediment discharge is written as:

$$q_t = q_b + \int_a^h C_a \frac{u_*}{\kappa} \left[\frac{h-z}{z} \frac{a}{h-a} \right]^{\omega/\beta_s \kappa u_*} \ln \frac{30z}{d_s} \, dz \tag{11.3}$$

The reference concentration $C_a = q_b/av_a$ is obtained from the unit bed sediment discharge q_b transported in the bed layer of thickness $a = 2d_s$, given the velocity v_a at the top of the bed layer, $v_a = (u_*/\kappa) \ln(30a/d_s) = 4.09u_*/\kappa$. Rewriting Equation (11.3) in dimensionless form with $z^* = z/h$, $E = 2d_s/h$, and $\mathrm{Ro} = \omega/\beta_s \kappa u_*$ gives

$$q_t = q_b \left[1 + I_1 \ln \frac{30h}{d_s} + I_2 \right] \tag{11.4}$$

where

$$I_1 = 0.216 \frac{E^{\mathrm{Ro}-1}}{(1-E)^{\mathrm{Ro}}} \int_E^1 \left[\frac{1-z^*}{z^*} \right]^{\mathrm{Ro}} dz^* \tag{11.5a}$$

$$I_2 = 0.216 \frac{E^{\mathrm{Ro}-1}}{(1-E)^{\mathrm{Ro}}} \int_E^1 \left[\frac{1-z^*}{z^*} \right]^{\mathrm{Ro}} \ln z^* \, dz^* \tag{11.5b}$$

The two integrals I_1 and I_2 can be solved numerically or with the use of nomographs prepared by Einstein. Einstein introduced several correction factors accounting for hydraulically smooth boundaries, sediment transport by size fraction, hiding factors, grain resistance instead of total resistance, velocity, and pressure correction. The reader is referred to the Appendix for the details of this complex method. This method is expected to work best when the bedload constitutes the most significant portion of the total load – thus, when $u_* < 2\omega$.

Einstein's procedure has been used to develop simpler methods such as those of Colby and Simons et al., described in Sections 11.1.2 and 11.1.3, respectively.

11.1.2   Colby's method

Colby (1964a,b) developed four graphical relations shown in Figures 11.2a and b for determining the bed sediment discharge. In arriving at his curves, Colby was guided by the Einstein bedload function (Einstein, 1950) and a large amount of data from streams and flumes. However, it should be mentioned that all curves for 100-ft depth, most curves of 10-ft depth, and some curves of 1.0 and 0.1 ft are not based on data but are extrapolated from limited data and theory.

In applying Figures 11.2a and b to the computation of the total bed sediment discharge, Colby's procedure is as follows: (1) The required data are the mean flow velocity V, the flow depth h, the median size of bed

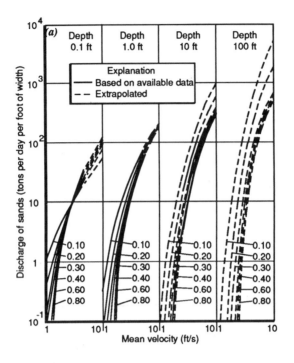

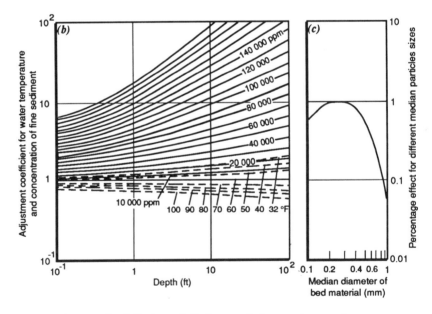

Figure 11.2. Colby's sand discharge relations and correction curves

material d_{50}, the water temperature $T°$, and the fine sediment concentration C_f. (2) The uncorrected sediment discharge q_{su} for the given V, h, and d_s can be found from Figure 11.2a by first reading q_{su} given values of V and d_s for two depths that bracket the desired depth and then interpolating on a logarithmic graph of depth versus q_{su} to get the bed sediment discharge per unit width. (3) The two correction factors c_{C1} and c_{C2} shown in Figure 11.2b account for the effect of water temperature and fine suspended sediment on the bed sediment discharge. If the bed sediment size falls outside the 0.20- to 0.60-mm range, the factor c_{C3} from Figure 11.2c is used to correct for the effect of sediment size. (4) The unit bed discharge q_{sc} corrected for the effect of water temperature, the presence of fine suspended sediment, and sediment size is given by the equation

$$q_{sc} = [1 + (c_{C1}c_{C2} - 1)c_{C3}]q_{su} \tag{11.6}$$

As Figure 11.2b indicates, $c_{C1} = 1$ when the temperature is 60°F, $c_{C2} = 1$ when the concentration of fine sediment is negligible, and $c_{C3} = 1$ when d_{50} lies between 0.2 and 0.3 mm. Case Study 11.1 shows bed sediment discharge calculations using Colby's method.

11.1.3 Simons, Li, and Fullerton's method

Simons, Li, and Fullerton (1981) developed an efficient method of evaluating sediment discharge. The method is based on easy-to-apply power relationships that estimate sediment transport based on the flow depth h and velocity V. These power relationships were developed from a computer solution of the Meyer-Peter and Müller bedload transport equation and Einstein's integration of the suspended bed sediment discharge:

$$q_s = c_{S1}h^{c_{S2}}V^{c_{S3}} \tag{11.7}$$

The results of the total bed sediment discharge are presented in Table 11.1. The large values of c_{S3} ($3.3 < c_{S3} < 3.9$) show the high level of dependence of sediment transport rates on velocity. Depth has comparatively less influence ($-0.34 < c_{S2} < 0.7$).

For flow conditions within the range outlined in Table 11.2, the regression equations should be accurate within 10%. The equations were obtained for steep sand- and gravel-bed channels under supercritical flow. They do not apply to cohesive material.

The equations assume that all sediment sizes are transported by the flow without armoring. Case Study 11.1 provides sample calculations of sediment transport using these power relationships.

Table 11.1. *Power equations for total bed sediment discharge in sand- and fine-gravel-bed streams*

$$q_s = c_{S1} h^{c_{S2}} V^{c_{S3}}$$

	d_{50} (mm)							
	0.1	0.25	0.5	1.0	2.0	3.0	4.0	5.0
Gr = 1.0								
c_{S1}	3.30×10^{-5}	1.42×10^{-5}	7.6×10^{-6}	5.62×10^{-6}	5.64×10^{-6}	6.32×10^{-6}	7.10×10^{-6}	7.78×10^{-6}
c_{S2}	0.715	0.495	0.28	0.06	−0.14	−0.24	−0.30	−0.34
c_{S3}	3.30	3.61	3.82	3.93	3.95	3.92	3.89	3.87
Gr = 2.0								
c_{S1}		1.59×10^{-5}	9.8×10^{-6}	6.94×10^{-6}	6.32×10^{-6}	6.62×10^{-6}	6.94×10^{-6}	
c_{S2}		0.51	0.33	0.12	−0.09	−0.196	−0.27	
c_{S3}		3.55	3.73	3.86	3.91	3.91	3.90	
Gr = 3.0								
c_{S1}			1.21×10^{-5}	9.14×10^{-6}	7.44×10^{-6}			
c_{S2}			0.36	0.18	−0.02			
c_{S3}			3.66	3.76	3.86			
Gr = 4.0								
c_{S1}				1.05×10^{-5}				
c_{S2}				0.21				
c_{S3}				3.71				

Definitions: q_s, unit sediment transport rate in ft²/s (unbulked); V, velocity in ft/s; h, depth in ft; Gr, gradation coefficient [Eq. (2.7b)].

Table 11.2. *Range of parameters for*
the Simons–Li–Fullerton method

Parameter	Value range
Froude number	1–4
Velocity	6.5–26 ft/s
Manning coefficient n	0.015–0.025
Bed slope	0.005–0.040
Unit discharge	10–200 ft/s
Particle size	$d_{50} \geq 0.062$ mm
	$d_{90} \leq 15$ mm

Case Study 11.1   Big Sand Creek, United States. A test reach of the Big
Sand Creek near Greenwood, Mississippi, has been used for bed sediment
discharge calculations by size fractions using the methods of Einstein, of
Colby, and of Simons, Li, and Fullerton. The sand-bed channel has a bed
slope $S_0 = 0.00105$. An average of four bed sediment samples is shown in
the following tabulation, showing that 95.8% of the bed material is between $d_s = 0.589$ mm and $d_s = 0.147$ mm, which is divided into four size
fractions for the calculations:

Grain size distribution (mm)	Average grain size			Settling velocity	
	mm	ft	Δp_i (%)	cm/sec	ft/s
$d_s > 0.60$	—	—	2.4	—	—
$0.60 > d_s > 0.42$	0.50	0.00162	17.8	6.2	0.205
$0.42 > d_s > 0.30$	0.36	0.00115	40.2	4.5	0.148
$0.30 > d_s > 0.21$	0.25	0.00081	32.0	3.2	0.106
$0.21 > d_s > 0.15$	0.18	0.00058	5.8	2.0	0.067
$0.15 > d_s$	—	—	1.8	—	—
$d_{16} = 0.24$ mm					
$d_{35} = 0.29$ mm $\sigma_g = 1.35$					
$d_{50} = 0.34$ mm					
$d_{65} = 0.37$ mm Gr $= 1.35$					
$d_{84} = 0.44$ mm					

Einstein's method. Calculations by size fractions using the Einstein method are detailed in the Appendix, given the cross-sectional geometry information from Figure A.7.

Colby's method. The water temperature and the fine sediment concentration for Colby's method are assumed equal to 70°F and 10,000 ppm, respectively. For convenience, the calculations are summarized in the form of tables. The following two tabulations show, respectively, the calculations by individual size fractions using the bed sediment size distribution and calculations over all size fractions using the median diameter of bed sediment:

d_{si}	Δp_i	h	W	V	q_{su}	c_{C1}	c_{C2}	c_{C3}	q_{sc}	$\Delta p_i q_{sc}$	$\Delta p_i Q_{sc}$
0.495	0.178	1.36	103	2.92	12	0.92	1.20	0.6	13	2.3	237
		1.76	136	4.44	40	0.91	1.21	0.6	43	7.7	1,050
		2.50	170	6.63	112	0.91	1.22	0.6	124	21.0	3,570
		3.30	194	8.40	193	0.90	1.23	0.6	213	37.0	7,180
		4.14	234	9.92	265	0.90	1.25	0.6	298	51.0	11,900
0.351	0.402	1.36	103	2.92	15	0.92	1.20	0.9	16	6.4	659
		1.76	136	4.44	45	0.91	1.21	0.9	49	20.0	2,720
		2.50	170	6.63	120	0.91	1.22	0.9	132	53.0	9,010
		3.30	194	8.40	210	0.90	1.23	0.9	230	93.0	18,000
		4.14	234	9.92	290	0.90	1.25	0.9	326	130.0	30,420
0.248	0.320	1.36	103	2.92	18	0.92	1.20	1.0	20	6.4	659
		1.76	136	4.44	53	0.91	1.21	1.0	58	19.0	2,580
		2.50	170	6.63	140	0.91	1.22	1.0	155	50.0	8,500
		3.30	194	8.40	240	0.90	1.23	1.0	266	85.0	16,500
		4.14	234	9.92	345	0.90	1.25	1.0	388	124.0	29,000
0.175	0.058	1.36	103	2.92	23	0.92	1.20	1.0	25	1.5	155
		1.76	136	4.44	64	0.91	1.21	1.0	70	4.1	558
		2.50	170	6.63	163	0.91	1.22	1.0	177	10.0	1,700
		3.30	194	8.40	305	0.90	1.23	1.0	331	20.0	3,880
		4.14	234	9.92	420	0.90	1.25	1.0	462	27.0	6,320

Definitions: d_{si}, given representative grain size for fraction i, mm; Δp_i, given fraction of bed material; h, given average flow depth, ft; W, given top width, ft; V, given average flow velocity, ft/s; q_{su}, uncorrected unit sediment discharge in tons/d × ft assuming that the bed is composed entirely of one sand size d_s, from Fig. 11.2a, by interpolation on logarithmic paper for the given V, h, and d_s; c_{C1}, correction factor for temperature, from Fig. 11.2b; c_{C2}, correction factor for fine sediment, from Fig. 11.2b; c_{C3}, correction factor for sediment size, from Fig. 11.2c; q_{sc}, corrected unit bed sediment discharge assuming that the bed is composed entirely of one sand size d_s, tons/d × ft width [Eq. (11.6)]; $\Delta p_i q_{sc}$, fraction weighted unit bed sediment discharge per unit width for a size fraction, tons/d × ft width; $\Delta p_i Q_{sc}$, $W_{iB} q_T$, fraction weighted bed sediment discharge, tons/d.

h	W	V	q_{su}	c_{C1}	c_{C2}	c_{C3}	q_{sc}	Q_{sc}
1.36	103	2.92	14.5	0.92	1.20	0.99	15.6	1,610
1.76	136	4.44	50.0	0.91	1.21	0.99	55.0	7,480
2.50	170	6.63	135.0	0.91	1.22	0.99	150.0	25,500
3.30	194	8.40	220.0	0.90	1.23	0.99	243.0	47,100
4.14	234	9.92	325.0	0.90	1.25	0.99	365.0	85,410

Definitions: h, given mean flow depth, ft; W, given surface width, ft; V, given average flow velocity, ft/s; q_{su}, uncorrected unit bed sediment discharge, tons/d × ft width, from Fig. 11.1a, for the given V, h, and $d_{50} = 0.34$ mm, by logarithmic interpolation; c_{C1}, correction factor for temperature, from Fig. 11.2b; c_{C2}, correction factor for fine sediment concentration, from Fig. 11.2b; c_{C3}, correction factor for sediment size, from Fig. 11.2b; q_{sc}, corrected unit bed sediment discharge in tons/d × ft given by Eq. (11.6); Q_{sc}, bed sediment discharge tons/d given by $Q_t = Wq_t$.

Simons, Li, and Fullerton's method. This method does not involve calculations by size fractions. However, the coefficients of the sediment transport equation can be interpolated from Table 11.1 given the gradation coefficient Gr and the median grain size d_{50}. In the case in point, $d_{50} = 0.34$ mm and Gr = 1.35, one obtains

$$q_s \,(\text{ft}^2/\text{s}) = 1.25 \times 10^{-5} h^{0.43} \,(\text{ft})\, V^{3.65} \,(\text{ft/s})$$

or

$$Q_s \,(\text{tons/ft} \times \text{d}) = 0.0893 W \,(\text{ft})\, h^{0.43} \,(\text{ft})\, V^{3.65} \,(\text{ft/s})$$

W (ft)	h (ft)	V (ft/s)	Q (ft³/s)	Q_s (tons/ft × d)
103	1.36	2.92	409	525
136	1.76	4.44	1,063	3,575
170	2.50	6.63	2,818	22,430
194	3.30	8.40	5,377	68,425
234	4.4	9.92	10,213	171,380

The results of bed sediment discharge calculations for Big Sand Creek using the methods of Einstein, Colby, and Simons, Li, and Fullerton are shown in Figure CS11.1.1. In all cases, the sediment discharge increases very rapidly with increasing water discharge.

11.1.4 Bagnold's method

Bagnold (1966) developed a sediment transport formula based on the concepts of energy balance. He stated that the available power of the flow supplies the energy for sediment transport. The resulting bed sediment

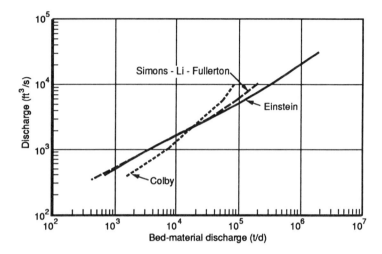

Figure CS11.1.1. Sample comparison of the methods of Einstein, Colby, and Simons, Li, and Fullerton

transport equation combines bedload and suspended load:

$$q_t = q_b + q_s = \frac{\tau_0 V}{G - 1}\left(e_B + 0.01\frac{V}{\omega}\right) \tag{11.8}$$

where $0.2 < e_B < 0.3$. The sediment discharge q_t is expressed as dry weight per unit time and width in any consistent system of units. Equation (11.8) is applicable to fully turbulent flows, and results are best for large transport rates.

Engelund and Hansen (1967) applied Bagnold's stream power concept and the similarity principle to obtain the sediment concentration by weight C_w as follows:

$$C_w = 0.05\left(\frac{G}{G - 1}\right)\frac{V S_f}{[(G - 1)g d_s]^{1/2}}\left[\frac{R_h S_f}{(G - 1)d_s}\right]^{0.5} \tag{11.9}$$

where d_s is the grain size, S_f the friction slope, R_h the hydraulic radius, V the depth-averaged velocity, g the gravitational acceleration, and G the specific gravity of sediment.

11.1.6   Ackers and White's method

Ackers and White (1973) postulated that only a part of the shear stress on the channel bed is effective in causing motion of coarse sediment. In the case of fine sediment, however, suspended load predominates and the total shear stress contributes to sediment motion. On this premise, the sediment mobility is described by the parameter

$$c_{AW5} = \frac{u_*^{c_{AW1}}}{\sqrt{(G-1)gd_s}} \left[\frac{V}{\sqrt{32} \log(10h/d_s)} \right]^{1-c_{AW1}} \tag{11.10}$$

in which c_{AW1} is zero for coarse sediment and unity for fine sediment. The total sediment concentration by weight is given by

$$C_w = c_{AW2} G \frac{d_s}{h} \left(\frac{V}{u_*} \right)^{c_{AW1}} \left[\frac{c_{AW5}}{c_{AW3}} - 1 \right]^{c_{AW4}} \tag{11.11}$$

in which c_{AW1}, c_{AW2}, c_{AW3}, and c_{AW4} depend on the dimensionless particle diameter $d_* = [(G-1)g/\nu^2]^{1/3}d_s$. The relationships for c_{AW1}, c_{AW2}, c_{AW3}, and c_{AW4} obtained using flume data for particle sizes ranging from 0.04 mm to 4.0 mm are

    1. for $1.0 < d_* < 60.0$,

$$c_{AW1} = 1.0 - 0.56 \log d_*$$

$$\log c_{AW2} = 2.86 \log d_* - (\log d_*)^2 - 3.53$$

$$c_{AW3} = \frac{0.23}{d_*^{1/2}} + 0.14$$

$$c_{AW4} = \frac{9.66}{d_*} + 1.34$$

    2. for $d_* > 60.0$,

$$c_{AW1} = 0, \quad c_{AW2} = 0.025, \quad c_{AW3} = 0.17, \quad c_{AW4} = 1.50$$

It can be seen that incipient motion occurs where $c_{AW3} = c_{AW5}$. Such a condition for incipient motion agrees well with Shields' criterion for coarse sediment, whereas for finer material, it gives results between those of Shields and White. Ackers and White's method tends to overestimate the concentration and sediment transport of fine and very fine sands.

11.1.7 Yang's method

Yang (1973) suggested that the total sediment concentration is related to potential energy dissipation per unit weight of water, that is, the unit stream power, which he expressed as the product of the velocity and slope. The dimensionless regression relationships for the total sediment concentration C_t in ppm by weight are

1. for sand,

$$\log C_{ppm} = 5.435 - 0.286 \log \frac{\omega d_s}{\nu} - 0.457 \log \frac{u_*}{\omega}$$

$$+ \left(1.799 - 0.409 \log \frac{\omega d_s}{\nu} - 0.314 \log \frac{u_*}{\omega}\right)$$

$$\times \log\left(\frac{VS}{\omega} - \frac{V_c S}{\omega}\right) \qquad (11.12a)$$

2. for gravel,

$$\log C_{ppm} = 6.681 - 0.633 \log \frac{\omega d_s}{\nu} - 4.816 \log \frac{u_*}{\omega}$$

$$+ \left(2.784 - 0.305 \log \frac{\omega d_s}{\nu} - 0.282 \log \frac{u_*}{\omega}\right)$$

$$\times \log\left(\frac{VS}{\omega} - \frac{V_c S}{\omega}\right) \qquad (11.12b)$$

in which the dimensionless critical velocity V_c/ω at incipient motion can be expressed as

$$\frac{V_c}{\omega} = \frac{2.5}{\log(u_* d_s/\nu) - 0.06} + 0.66 \quad \text{for } 1.2 < \frac{u_* d_s}{\nu} < 70 \qquad (11.13a)$$

and

$$\frac{V_c}{\omega} = 2.05 \quad \text{for } \frac{u_* d_s}{\nu} \geq 70 \qquad (11.13b)$$

These equations are dimensionless, including the total sediment concentration C_{ppm} in ppm by weight, V_c is the average flow velocity at incipient motion, VS is the unit stream power, and VS/ω is the dimensionless unit stream power. Flume and field data range from 0.137 to 1.71 mm for particle size, and 0.037 to 49.9 ft for water depth. However, the majority of the data cover medium to coarse sands at flow depths rarely exceeding 3 ft.

11.1.8   Shen and Hung's method

Shen and Hung (1972) recommended a regression formula based on available data for engineering analysis of sediment transport. They selected the sediment concentration as the dependent variable and the fall velocity ω in ft/s of the median diameter of bed material, flow velocity V in ft/s, and energy slope as independent variables. The concentration of bed sediment by weight in ppm is given as a power series of the flow parameter, based on 587 data points:

$$\log C_{ppm} = [-107,404.459 + 324,214.747 \, \text{Sh} - 326,309.589 \, \text{Sh}^2$$
$$+ 109,503.872 \, \text{Sh}^3] \qquad (11.14a)$$

where

$$\text{Sh} = \left[\frac{VS^{0.57159}}{\omega^{0.31988}} \right]^{0.00750189} \qquad (11.14b)$$

The fall velocity of sediment particles was corrected to the actual measured water temperature but not to include the effect of significant concentrations of fine sediment on bed-material transport. It is important not to round off the coefficients and exponents of Equations (11.14a) and (11.14b). This formula performs quite well with flume data, but tends to underpredict the total load of large rivers like the Rio Grande, the Mississippi, the Atchafalaya, the Red River, and some large canals in Pakistan.

11.1.9 Brownlie's method

Brownlie (1981) obtained the following equation for the concentration C_{ppm}:

$$C_{ppm} = 7115 c_B \left(\frac{V - V_c}{\sqrt{(G-1)gd_s}} \right)^{1.978} S_f^{0.6601} \left(\frac{R_h}{d_s} \right)^{-0.3301} \qquad (11.15a)$$

in which the value of V_c is given in terms of the Shields dimensionless critical shear stress τ_{*c}, the friction slope S_f, and the geometric standard deviation of the bed material σ_g by the equation

$$\frac{V_c}{\sqrt{(G-1)gd_s}} = 4.596 \tau_{*c}^{0.529} S_f^{-0.1405} \sigma_g^{-0.1606} \qquad (11.15b)$$

The coefficient c_B is unity for laboratory data and 1.268 for field data.

11.1.10 Karim and Kennedy's method

Karim and Kennedy (1981) carried out a regression analysis of the sediment data from laboratory flumes and natural streams,

$$\log \frac{q_t}{\gamma_s \sqrt{(G-1)gd_s^3}} = -2.28 + 2.97 c_{k1} + 0.30 c_{k2} c_{k3} + 1.06 c_{k1} c_{k3} \qquad (11.16a)$$

where

$$c_{k1} = \log \frac{V}{\sqrt{(G-1)gd_s}}; \qquad c_{k2} = \log\left(\frac{h}{d_s}\right);$$

$$c_{k3} = \log\left(\frac{u_* - u_{*c}}{\sqrt{(G-1)gd_s}}\right)$$

(11.16b)

Karim and Kennedy found good predictions for a large amount of flume and field data. The equations of Brownlie and of Karim and Kennedy deserve further testing.

Case Study 11.2 compares sediment transport calculations on the Colorado River by the methods of Bagnold, Engelund and Hansen, Ackers and White, Yang, Shen and Hung, Karim and Kennedy, and Brownlie.

Case Study 11.2   Colorado River, United States. The Colorado River at Taylor's Ferry carries a significant volume of sand. The bed-material size has a geometric mean of 0.32 mm and standard deviation $\sigma_g = 1.44$, $d_{35} = 0.287$ mm, $d_{50} = 0.33$ mm, $d_{65} = 0.378$ mm. The detailed sieve analysis is given in the following tabulation:

Sieve opening (mm)	% Finer
0.062	0.22
0.125	1.33
0.250	21.4
0.5	88.7
1.0	98.0
2.0	99.0
4.0	99.5

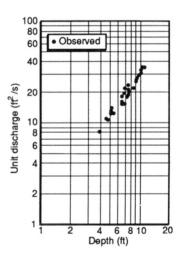

The stage–discharge relationship for the Colorado River is shown in the above figure at flow depths ranging from 4 to 12 ft and unit discharges from 8 to 35 ft²/s. Consider the channel slope 0.000217, channel width 350 ft, and a water temperature of 60°F. Calculate the unit sediment discharge by size fraction using the methods detailed in Sections 11.1.4 to 11.1.10. Plot the results on the given sediment rating curve shown in Figure CS11.2.1.

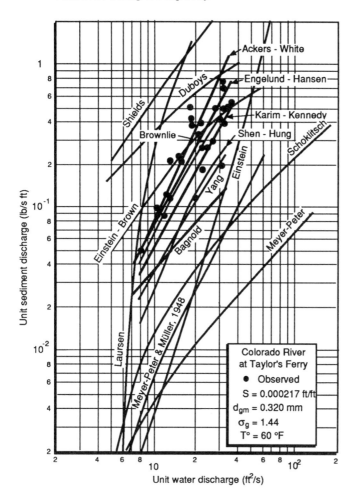

Figure CS11.2.1. Sediment-rating curve of the Colorado River
(after Vanoni et al., 1960)

For the sake of comparison of several sediment transport equations,
flow depths (a) $h = 10$ ft and (b) $h = 4$ ft have been selected, respectively.
At each flow depth, several sediment transport equations are compared
in terms of potential concentration in ppm for each size fraction, and
then the fraction-weighted sediment discharge in lb/ft × s is given in the
second tabulation.

Note that from Equation (10.1),

$$C_{\text{ppm}} = 10^6 C_w \quad \text{and} \quad C_{\text{mg/l}} = \frac{GC_{\text{ppm}}}{G + (1-G)C_w}$$

The unit sediment discharge q_s in N/m $\times$ s is obtained from

$$q_s \, (\text{N/m} \times \text{s}) = 10^{-3} g \, (\text{m/s}^2) \, C \, (\text{mg/l}) \, q \, (\text{m}^2/\text{s})$$

$$q_s \, (\text{lb/ft} \times \text{s}) = q_s \, (\text{N/m} \times \text{s}) \times \frac{\text{lb}}{4.45 \, \text{N}} \times \frac{\text{m}}{3.28 \, \text{ft}} = 0.0685 q_s \, (\text{N/m} \times \text{s})$$

(a)				
Flow depth	10 ft	or	3.05 m	
Velocity	3.2 ft/s	or	0.97 m/s	
Unit discharge	32 ft²/s	or	2.97 m²/s	
Hydraulic radius	9.45 ft	or	2.9 m	
Shear stress	0.128 lb/ft²	or	6.14 N/m²	
Shear velocity	0.257 ft/s	or	0.078 m/s	
Water temperature	60°F	or	15.4°C	
Kinematic viscosity	1.21×10^{-5} ft²/s	or	1.12×10^{-6} m²/s	

Δp_i	d_s (mm)	d_*	ω (m/s)	u_*/ω	τ_c (N/m²)
0.002	0.042	0.97	0.001	56.7	—
0.011	0.083	1.95	0.005	14.5	0.166
0.201	0.167	3.89	0.019	4.1	0.166
0.673	0.333	7.79	0.049	1.65	0.203
0.093	0.667	15.5	0.085	0.92	0.328
0.010	1.33	31.1	0.132	0.59	0.761
0.010	2.66	62.3	0.193	0.41	1.86

d_s	Bagnold	Engelund–Hansen	Ackers–White	Yang	Shen–Hung	Karim–Kennedy	Brownlie
Potential sediment concentration C (ppm)							
0.042	960	5,959	?[a]	3,529	1,439	7,584	1,133
0.083	270	2,106	27,317	567	586	1,289	491
0.167	101	745	584	162	222	454	313
0.333	60	263	140	93	97	181	191
0.667	48	93	69	91	54	79	107
1.33	43	33	31	113	34	35	48
2.67	40	12	5.9	1.8	22	14	14
Fraction weighted sediment transport $\Delta p_i q_s$ (N/m$\times$s)							
0.042	0.06	0.38	?[a]	0.23	0.09	0.49	0.07
0.083	0.09	0.68	9.0	0.18	0.19	0.42	0.16
0.167	0.59	4.36	3.42	0.95	1.30	2.66	1.84
0.333	1.18	5.17	2.75	1.83	1.92	3.55	3.76
0.667	0.13	0.25	0.019	0.25	0.15	0.22	0.29
1.33	0.01	0.01	0.01	0.03	0.01	0.01	0.01
2.67	0.01	0	0	0	0.01	0	0

d_s	Bagnold	Engelund–Hansen	Ackers–White	Yang	Shen–Hung	Karim–Kennedy	Brownlie
Total							
N/m × s	2.07	10.8	15.37?[a]	3.48	3.66	7.35	6.13
lb/ft × s	0.142	0.74	1.05	0.234	0.25	0.50	0.42

[a] Extremely high.

(b)

Flow depth	4 ft	or	1.22 m
Velocity	2 ft/s	or	0.61 m/s
Unit discharge	8 ft²/s	or	0.74 m²/s
Hydraulic radius	3.91 ft	or	1.2 m
Shear stress	0.053 lb/ft²	or	2.538 N/m²
Shear velocity	0.1652 ft/s	or	0.05 m/s
Water temperature	60°F	or	15.4°C
Kinematic viscosity	1.21×10^{-5} ft²/s	or	1.12×10^{-6} m²/s

Δp_i	d_s (mm)	d_*	ω (m/s)	u_*/ω	τ_c (N/m²)
0.002	0.042	0.974	0.001	36.44	—
0.011	0.083	1.947	0.005	9.3	0.166
0.201	0.167	3.894	0.019	2.7	0.166
0.673	0.333	7.788	0.047	1.06	0.203
0.093	0.667	15.577	0.085	0.59	0.328
0.010	1.333	31.153	0.132	0.38	0.761
0.010	2.667	62.307	0.193	0.26	1.86

d_s	Bagnold	Engelund–Hansen	Ackers–White	Yang	Shen–Hung	Karim–Kennedy	Brownlie
Potential sediment concentration C (ppm)							
0.042	612	1,540	?[a]	1,570	543	1,522	598
0.083	181	544	2,017	250	184	255	198
0.167	75	192	142	69	56	116	127
0.333	50	68	57	37	21	58	75
0.667	42	24	30	34	10	31	38
1.33	39	8.5	5.2	42	5.7	14	11.1
2.67	37	3.0	0	1.2	3.4	3.3	0.03
Fraction weighted sediment transport $\Delta p_i q_s$ (N/m × s)							
0.042	0.01	0.02	?[a]	0.03	0.01	0.02	0.01
0.083	0.01	0.04	0.16	0.02	0.01	0.02	0.02
0.167	0.11	0.28	0.21	0.10	0.08	0.17	0.19
0.333	0.24	0.33	0.28	0.18	0.10	0.29	0.37

d_s	Bagnold	Engelund–Hansen	Ackers–White	Yang	Shen–Hung	Karim–Kennedy	Brownlie
0.667	0.03	0.02	0.02	0.02	0.01	0.02	0.03
1.33	0	0	0	0	0	0	0
2.67	0	0	0	0	0	0	0
Total							
N/m×s	0.41	0.70	0.67?[a]	0.36	0.21	0.52	0.61
lb/ft×s	0.0283	0.0481	0.0459	0.0245	0.0147	0.0359	0.0417

[a] Extremely high.

11.2   Supply-limited sediment transport

In streams with very coarse bed material, stiff clay, or bedrock control, the sediment transport capacity of fine fractions calculated from sediment transport formulas far exceeds the sediment supply from upstream sources. Sediment transport in such streams is limited by the upstream supply of sediments, and sediment transport estimates can be obtained from an analysis of sediment sources (Section 11.2.1) and sediment yield (Section 11.2.2).

11.2.1   Sediment sources

The analysis of sediment sources aims at estimating the total amount of sediment eroded on the watershed on an annual basis, called *annual gross erosion*. The annual gross erosion A_T depends on the source of sediments in terms of upland erosion A_U, gully erosion A_G, and local bank erosion A_B; thus, $A_T = A_U + A_G + A_B$.

Upland erosion A_U generally constitutes the primary source of sediment; other sources of gross erosion, such as mass wasting or bank erosion A_B and gully erosion A_G, must be estimated separately by calculating the annual volume of sediment scoured through lateral migration of the stream and the upstream migration of headcuts. In stable fluvial systems, the analysis of sediment sources focuses on upland erosion losses from rainfall and snowmelt.

The impact of raindrops on a soil surface can exert a surface shear stress up to 80 Pa, thus far exceeding the bonding forces between soil particles (Hartley and Julien, 1992). The detached particles are transported through sheet flow into rills and small channels. With reference to Example 2.2, the unit upland sediment discharge from sheet and rill erosion can be written in the form

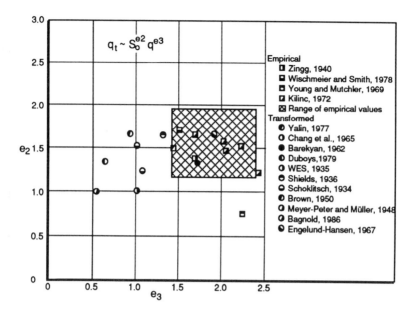

Figure 11.3. Exponents of the sediment transport equation for sheet flow

$$q_t = \tilde{e}_1 S_0^{e_2} q^{e_3} \tag{11.17a}$$

The values of the exponents e_2 and e_3 from field observations and from bedload equations are shown in Figure 11.3. The typical range for field values is $1.2 < e_2 < 1.9$ and $1.4 < e_3 < 2.4$. The equation of Kilinc (1972) for sheet and rill erosion is recommended for bare sandy soils:

$$q_t \, (\text{lb/ft} \times \text{s}) = 1.24 \times 10^5 S_0^{1.66} q^{2.035} \qquad (q \text{ in ft}^2/\text{s}) \tag{11.17b}$$

$$q_t \, (\text{tons/m} \times \text{s}) = 2.55 \times 10^4 S_0^{1.66} q^{2.035} \qquad (q \text{ in m}^2/\text{s}) \tag{11.17c}$$

Considering various soil types and vegetation, the annual rainfall erosion losses A_U can be calculated from the universal soil-loss equation (USLE). The USLE computes soil losses at a given site from the product of six major factors:

$$A_U = \hat{R}\hat{K}\hat{L}\hat{S}\hat{C}\hat{P} \tag{11.18}$$

where A_U is the soil loss per unit area from sheet and rill erosion, normally in tons/acre, $\hat{R}$ the rainfall erosivity factor, $\hat{K}$ the soil erodibility factor, usually in tons/acre, $\hat{L}$ the field length factor, $\hat{S}$ the field slope factor, $\hat{C}$ the cropping-management factor normalized to a tilled area with continuous fallow, and $\hat{P}$ the conservation practice factor normalized to straight-row farming up and down the slope.

The rainfall erosivity factor $\hat{R}$ can be calculated for each storm from

$$\hat{R} = 0.01 \, \Sigma(916 + 331 \log I)I \tag{11.19}$$

where I is the rainfall intensity in in./h. The annual rainfall erosion index in the United States decreases from a value exceeding 500 near the Gulf of Mexico to values under 100 in the northern states and in the Rockies. The slope length–steepness factor $\hat{L}\hat{S}$ is a topographic factor that can be approximated from the field runoff length X_r in ft and surface slope S_0 in ft/ft by

$$\hat{L}\hat{S} = \sqrt{X_r}\,(0.0076 + 0.53S_0 + 7.6S_0^2) \tag{11.20}$$

The factor $\hat{L}\hat{S}$ is normalized to be a runoff length of 72.6 ft and a field slope of 9%.

In the more general case of erosion from sheet flow, modifications to Equation (11.17) reflect the influence of soil type, vegetation, and practice factors using factors $\hat{K}$, $\hat{C}$, and $\hat{P}$ as

$$q_t \text{ (tons/m} \times \text{s)} = 1.7 \times 10^5 S_0^{1.66} q^{2.035} \hat{K}\hat{C}\hat{P} \tag{11.21a}$$

where the surface slope S_0 is in m/m and the unit discharge q is in m²/s. The dimensionless soil erodibility factor $\hat{K}$, the cropping-management factor $\hat{C}$, and the conservation practice factor $\hat{P}$ are obtained from Tables 11.3, 11.4, and 11.5, respectively. Whereas Equation (11.18) is limited to rainfall erosion losses, Equation (11.21) is applicable to both rainfall and snowmelt erosion losses.

The equivalent upland erosion is then calculated from

$$A_u = \int_{\text{time}} \int_{\text{width}} q_t \, dw \, dt \tag{11.21b}$$

11.2.2   Sediment yield

The rate at which sediment is carried by natural streams is much lower than the gross erosion on its upstream watershed. Sediment is deposited between the source and the stream cross section whenever the transport capacity of runoff water is insufficient to sustain transport.

The sediment delivery ratio S_{DR} denotes the ratio of the sediment yield Y at a given stream cross section to the gross erosion A_T from the watershed upstream from the measuring point. The sediment yield can therefore be written as

$$Y = A_T S_{DR} \tag{11.22}$$

Table 11.3. *Soil erodibility
factor $\hat{K}$ (tons/acre)*

Textural class	Organic matter content (%)	
	0.5	2
Fine sand	0.16	0.14
Very fine sand	0.42	0.36
Loamy sand	0.12	0.10
Loamy very fine sand	0.44	0.38
Sandy loam	0.27	0.24
Very fine sandy loam	0.47	0.41
Silt loam	0.48	0.42
Clay loam	0.28	0.25
Silty clay loam	0.37	0.32
Silty clay	0.25	0.23

Source: After Schwab et al. (1981).

Table 11.4a. *Cropping management factor $\hat{C}$
for undisturbed forest land*

Percentage of area covered by canopy of trees and undergrowth	Percentage of area covered by duff at least 2 in. deep	Factor $\hat{C}$
100–75	100–90	0.0001–0.001
70–45	85–75	0.002–0.004
40–20	70–40	0.003–0.009

Source: After Wischmeier and Smith (1978).

The sediment–delivery ratio depends primarily on the drainage area A_t of the upstream watershed. Values for the sediment–delivery ratio are given in Figure 11.4.

11.3   Sediment-rating curves

Sediment-rating curves display the rate of sediment transport as a function of the flow discharge. The rate of sediment transport is given in terms of sediment discharge or, alternatively, as a flux-averaged concentration. The analysis of sediment-rating curves depends on whether sediment transport

Table 11.4b. *Cropping management factor $\hat{C}$ for permanent pasture, range, and idle land*

		Percent ground cover					
Vegetative canopy[a]	Type[b]	0	20	40	60	80	95+
No appreciable canopy	G	0.45	0.20	0.10	0.042	0.013	0.003
	W	0.45	0.24	0.15	0.091	0.043	0.011
Tall weeds or short brush with average drop fall height of 20 in.	G	0.17–0.36	0.10–0.17	0.06–0.09	0.032–0.038	0.011–0.013	0.003
	W	0.17–0.36	0.12–0.20	0.09–0.13	0.068–0.083	0.038–0.042	0.011
Appreciable brush or bushes, with average drop fall height of 6.5 ft	G	0.28–0.40	0.14–0.18	0.08–0.09	0.036–0.040	0.012–0.013	0.003
	W	0.28–0.40	0.17–0.22	0.12–0.14	0.078–0.087	0.040–0.042	0.011
Trees, but not appreciable low brush; average drop fall height of 13 ft	G	0.36–0.42	0.17–0.19	0.09–0.10	0.039–0.041	0.012–0.013	0.003
	W	0.36–0.42	0.20–0.23	0.13–0.14	0.084–0.089	0.041–0.042	0.011

Note: The listed $\hat{C}$ values assume that the vegetation and mulch are randomly distributed over the entire area.

[a] Canopy height is measured as the average fall height of water drops falling from the canopy to the ground. The canopy effect is inversely proportional to drop fall height and is negligible if the fall height exceeds 33 ft.

[b] G: cover at surface is grass, grasslike plants, decaying compacted duff, or litter at least 2 in. deep. W: cover at surface is mostly broadleaf herbaceous plants (as weeds with little lateral-root network near the surface) or undecayed residues or both.

Source: Modified from Wischmeier and Smith (1978).

Table 11.4c. *Cropping management factor*
Ĉ for construction slopes

Type of mulch	Mulch rate (tons/acre)	Factor Ĉ
Straw	1.0–2.0	0.06–0.20
Crushed stone, 0.25–1.5 in.	135	0.05
	240	0.02
Wood chips	7	0.08
	12	0.05
	25	0.02

Source: Modified from Wischmeier and Smith (1978).

Table 11.5. *Conservation practice factor P̂ for*
contouring, strip cropping, and terracing

Land slope (%)	Farming on contour	Contour strip crop	Terracing (a)	Terracing (b)
2–7	0.50	0.25	0.50	0.10
8–12	0.60	0.30	0.60	0.12
13–18	0.80	0.40	0.80	0.16
19–24	0.90	0.45	0.90	0.18

(a) For erosion-control planning on farmland.
(b) For prediction of contribution to off-field sediment load.
Source: After Wischmeier (1972).

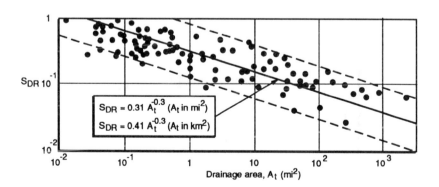

$$S_{DR} = 0.31 \, A_t^{-0.3} \; (A_t \text{ in mi}^2)$$
$$S_{DR} = 0.41 \, A_t^{-0.3} \; (A_t \text{ in km}^2)$$

Figure 11.4. Sediment–delivery ratio (after Boyce, 1975)

is limited by the sediment transport capacity of the stream or the up-stream supply of sediment. Section 11.3.1 covers the analysis of bed sediment transport capacity, whereas Section 11.3.2 deals with supply-limited sediment-rating curves.

11.3.1   Capacity-limited sediment-rating curves

When sediment transport is controlled by the transport capacity of bed sediment, the analysis of sediment-rating curves can be considered in two ways:

1. comparative analysis of sediment transport capacity curves by size fractions; or
2. comparative analysis of sediment transport formulas with measured field data.

The analysis of sediment transport capacity curves by size fractions involves plotting the sediment transport capacity from various formulas versus the particle size. An example is shown in Figure 11.5 for four sediment transport formulas, at two flow depths, 2 and 10 ft, respectively, all other conditions being identical.

The transport capacity increases largely with flow depth and varies inversely with sand grain size. In this example from Williams and Julien

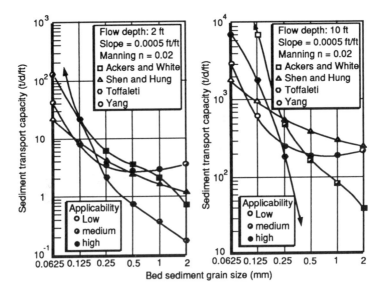

Figure 11.5. Typical sediment transport capacity curves

(1989), an applicability index has been defined as the number of parameters within the range of experimental data under which a given sediment transport formula was derived. Comparison of these four sediment transport formulas highlights a number of interesting features:

1. the agreement between different formulas is generally better when the applicability index is high;
2. the transport rates calculated using the methods of Ackers and White (1973) and of Toffaleti (1968) are very high for fine sands and very low for coarse sands; and
3. Yang's (1973) method shows a slight increase in sediment transport capacity with grain size for coarse sands.

In general, the transport capacity calculated by size fraction exceeds the transport capacity calculations based on the median grain size. For instance, consider a sand size distribution of 33% at 0.125 mm, 33% at 0.25 mm, and 33% at 0.5 mm. At a flow depth of 10 ft, transport capacity calculations by size fractions using Ackers and White's equation in Figure 11.5 give $0.33 \times 7,000$ tons/ft $\times$ d $+ 0.33 \times 500$ tons/ft $\times$ d $+ 0.33 \times 180$ tons/ft $\times$ d $= 2,560$ tons/ft $\times$ d, which far exceeds calculations based on the median grain size, 1×500 tons/ft $\times$ d $= 500$ tons/ft $\times$ d.

A comparison with field sediment discharge measurements is essential in the analysis of sediment-rating curves. The sediment transport rates by several formulas can vary by orders of magnitude. In many cases, it is virtually impossible to determine which sediment transport capacity formula should be used without valuable field measurements of sediment discharge.

The bed sediment-rating curve in Figure 11.6 for the Niobrara River near Cody, Nebraska, illustrates the variability in calculated sediment discharge from various formulas at a given discharge. Case Study 11.2 also provides a detailed analysis.

Relatively little scatter in field measurements is observed when the sediment load is controlled by the sediment transport capacity. In such cases, variations in sediment-rating curves are due to variability in water temperature, stream slope, bed sediment size, particle size distribution, and measurement errors. For capacity-limited sediment transport, the sediment-rating curve often fits a power law of the form

$$q_s = \bar{a} q^{\bar{b}} \tag{11.23}$$

Fitting straight lines on logarithmic graphs within the range of observed discharge often gives an exponent $\bar{b}$ ranging between 2 and 3. The flux-

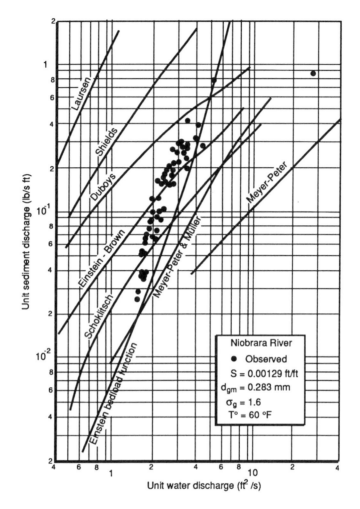

Figure 11.6. Sediment-rating curve of the Niobrara River (after Vanoni et al., 1960)

averaged sediment concentration C_f is given by the ratio of sediment discharge to water discharge. From Equation (11.23) this concentration is

$$C_f = \frac{q_s}{q} = \bar{a} q^{\bar{b}-1} \tag{11.24}$$

It is reasonable to expect a concentration increase during floods as long as the exponent $\bar{b}$ exceeds unity. The determination of coefficients $\bar{a}$ and $\bar{b}$ by regression analysis should be based on Equation (11.24), as opposed to Equation (11.23), to avoid spurious correlation.

11.3.2 Supply-limited sediment-rating curves

The case of supply-limited rating curves is characterized by low concentrations and high variability. Sediment transport is limited by the supply of sediment, usually washload, which varies with the location and intensity of rainstorms on the watershed (forest vs. agricultural fields), seasonal variation in temperature, weathering, vegetation, and type of precipitation (rain or snow). The source of sediment includes upland erosion, streambank erosion, point sources, and snowmelt. The example of the Bell River in Figure 11.7 is typical of supply-limited sediment transport with low concentration and large variability.

When the sediment concentration data are widely dispersed around the sediment curve obtained by regression analysis, better results are obtained by subdividing the discharge into small intervals and taking the average value (without logarithmic transformation) of sediment concentration for each interval. The sediment-rating curve is then hand-plotted from these mean values of the concentration. This procedure circumvents the bias introduced by linear regression analysis of log-transformed variables. This procedure also avoids the problems of mathematically fitting straight lines through curvilinear sediment-rating relationships.

Hysteresis effects between discharge and concentration, seasonal variation, inaccuracies in flow and sediment measurements, and variability in the washload may explain the scatter of points on the sediment transport graph. Better results are sometimes achieved, provided that sufficient data are available, by setting individual sediment-rating curves for each month. At a given discharge, higher sediment concentrations are generally observed during the rising limb of the hydrograph (Fig. 11.8). Different sediment-rating curves for the rising and falling limbs of hydrographs can occasionally be identified (Fig. 11.7). Sediment supply from streambank erosion can sometimes be separated from upland sediment sources. The origin of sed-

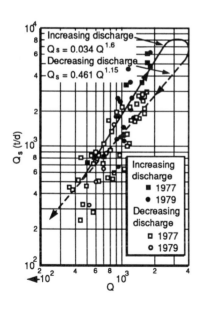

Figure 11.7. Sediment-rating curves of the Bell River (after Frenette and Julien, 1987)

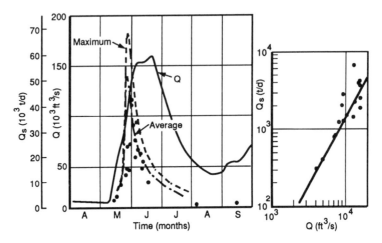

Figure 11.8. Suspended load in La Grande River
(after Frenette and Julien, 1987)

iment can be identified when sediments from different sources have different mineralogy, clay content, percentage of organic matter, color, concentration, and water chemistry. At times, the name of the river alone provides an indication of the type of sediment transport – for example, Muddy Creek, Red River, Green River, Colorado River, Black River, White River, Chalk Creek, Caine River, Platte River. Finally, in northern streams, snowmelt erosion rates can be compared with rainfall erosion rates (e.g., Case Study 11.3).

11.4   Short- and long-term sediment load

The short-term analysis of sediment load in Section 11.4.1 provides information, generally on a daily basis, on the magnitude and variability of sediment transport during rainstorm or snowmelt events. The long-term analysis, in contrast, gives an estimate of the expected amount of sediment yielded by a stream. On an annual basis, it gives the mean annual sediment load of a stream, covered in Section 11.4.2. The long-term sediment load is required for reservoir sedimentation, sediment budget, and specific degradation studies.

11.4.1   Daily sediment load

Daily sediment discharge can be computed with a relatively high degree of accuracy when the discharge and sediment concentration do not change rapidly. The total sediment discharge in tons/d is the product of the flux-averaged total sediment concentration, the daily mean water discharge, and a conversion factor. The daily sediment load is obtained by one of the following formulas:

$$Q_s \text{ (metric tons/d)} = 0.0864 C_{mg/l} Q \text{ (m}^3/\text{s)} \tag{11.25a}$$

or

$$Q_s \text{ (metric tons/d)} = 2.446 \times 10^{-3} C_{mg/l} Q \text{ (ft}^3/\text{s)} \tag{11.25b}$$

During periods of rapidly changing concentration and water discharge, the concentration and gage height graphs are subdivided into hourly time increments. Incomplete sediment records in which daily discharge measurements are sparse can be analyzed by first obtaining the sediment-rating curve from the measurements using the method of Section 11.3.2. Alternatively, nonlinear regression of log-transformed concentration versus discharge measurements is also possible. The missing sediment concentration can then be reconstituted from discharge measurements and the sediment-rating curve. A typical graph of daily sediment discharge for the York River is shown in Figure 11.9.

11.4.2 Annual sediment load

There are two basic approaches to the determination of the long-term average sediment load of a river. First, the summation over a long period of time of the measured and reconstituted daily sediment discharges from Section 11.4.1 can sometimes be accomplished using computers. The second approach uses both a sediment-rating curve between total sediment discharge, or flux-averaged concentration, and water discharge; and a

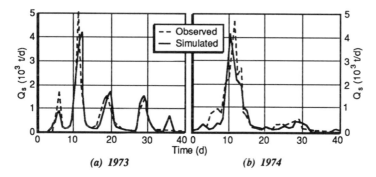

Figure 11.9. Daily sediment load simulation of the York River (after Frenette and Julien, 1987)

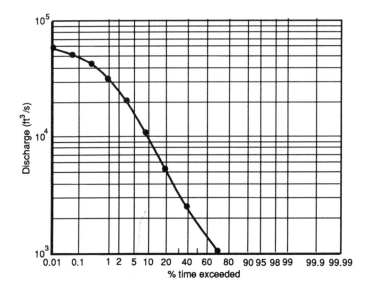

Figure 11.10. Flow-duration curve of the Chaudière River

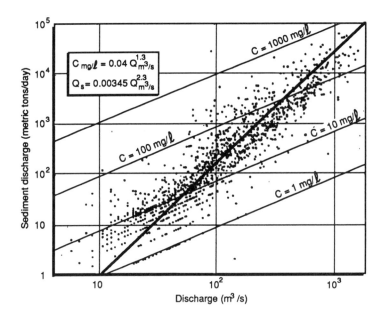

Figure 11.11. Sediment-rating curve of the Chaudière River

Table 11.6. *Long-term sediment yield of the Chaudière River from the flow-duration sediment-rating-curve method*

Time intervals (%) (1)	Interval midpoint (%) (2)	Interval Δp (%) (3)	Discharge Q_{cfs} (ft³/s) (4)	Concentration C (mg/l)[a] (5)	$Q_{cfs} \times \Delta p$ (ft³/s) (6)	Sediment load $Q_s \times \Delta p$ (tons/year)[b] (7)
0.00–0.02	0.01	0.02	58,000	607	12	6,280
0.02–0.1	0.06	0.08	52,000	526	42	19,539
0.1–0.5	0.3	0.4	43,000	411	172	63,102
0.5–1.5	1.0	1.0	33,000	292	330	85,820
1.5–5.0	3.25	3.5	21,000	162	737	106,213
5–15	10	10	10,640	67	1,064	63,529
15–25	20	10	5,475	29	548	13,782
25–35	30	10	3,484	16	348	4,873
35–45	40	10	2,435	10	244	2,138
45–55	50	10	1,839	7	184	1,121
55–65	60	10	1,375	4.7	138	575
65–75	70	10	1,030	3.2	103	296
75–85	80	10	763	2.1	76	149
85–95	90	10	547	1.4	55	69
95–98.5	96.75	3.5	397	0.9	14	12
Total						367,500

Notes: Columns (2) and (4) define the flow-duration curve. Columns (4) and (5) define the sediment-rating curve. The product of Columns (3) and (4) is given in Column (6).
[a] The concentration C in mg/l is calculated from $C_{mg/l} = 0.04 Q_{cms}^{1.3} = 3.89 \times 10^{-4} Q_{cfs}^{1.3}$.
[b] The annual sediment yield in metric tons/year is calculated from $Q_s \times \Delta p = 0.893 \times C \times Q \times \Delta p$ with C in mg/l and Q in ft³/s.

flow-duration curve. This method is referred to as the flow-duration/sediment-rating-curve method.

The flow-duration curve indicates the percentage of time a given river discharge is exceeded. As an example, the flow-duration curve of the Chaudière River is plotted on Figure 11.10, and discrete values are reported in columns (2) and (4) of Table 11.6. Notice that the selected time percent intervals are smaller as discharge increases. The sediment-rating curve of the Chaudière River at St.-Lambert-de-Lévis is shown in Figure 11.11, from which the flux-averaged concentration is approximated by $C_{mg/l} = 3.88 \times 10^{-4} Q^{1.3}$ ft³/s. Sediment concentration calculations for each interval are given in Table 11.6, column (5).

Experience indicates that the flow-duration/sediment-rating-curve method is most reliable (1) when the recording period is long, (2) when sufficient data at high flows are available, and (3) when the sediment-rating curve shows considerable scatter. Flood flows carry most of the sediments, and in order to give full weight to the total sediment inflow, field programs should cover the flood period. Unfortunately, measurements are rarely available during extreme events.

On an annual basis, the sediment load is given by

$$Q_s \text{ (metric tons/year)} = 31.56 C_{mg/l} Q \text{ (m}^3/\text{s)} \qquad (11.26a)$$

or

$$Q_s \text{ (metric tons/year)} = 0.893 C_{mg/l} Q \text{ (ft}^3/\text{s)} \qquad (11.26b)$$

The total annual sediment load is then given by the sum of all the intervals of the flow-duration curve. In this example, the sum of all numbers in column (7) of Table 11.6 gives an average annual sediment load of 367,500 metric tons/year for the Chaudière River. Case Study 11.3 illustrates the application of several concepts of supply-limited sediment transport to the Chaudière watershed in Canada.

Case Study 11.3   Chaudière River, Canada. The Chaudière River drains a 5,830-km^2 Appalachian basin to the St. Lawrence River near Quebec (Figures CS11.3.1a and b). Approximately 65% of the watershed area is still forested, whereas 35% supports agriculture and pasture (Figure CS11.3.1c). An analysis of soil erosion losses using the USLE and Equation (11.21) shows that soil erosion losses are primarily a function of the surface slope S_0 and the crop-management factor $\hat{C}$ of the USLE. The mean annual gross rainfall erosion from upland sources calculated from the USLE is of the order of 4.5×10^6 metric tons/year (Figure CS11.3.1d).

Short-term simulations of sediment discharge in Figures CS11.3.2 are possible despite the large scatter on the sediment-rating curve in Figure 11.11. Monthly simulations of both rainfall and snowmelt erosion losses in Figure CS11.3.3 demonstrate that about 70% of the mean annual sediment load results from snowmelt. The long-term average sediment load is estimated in Table 11.6.

Figure CS11.3.1. (a) Location of the Chaudière watershed (after Julien and Frenette, 1987); (b) surface slope of the Chaudière watershed (after Frenette and Julien, 1986b); (c) land use of the Chaudière watershed (after Frenette and Julien, 1986b); (d) annual soil erosion losses in metric tons (after Julien, 1979)

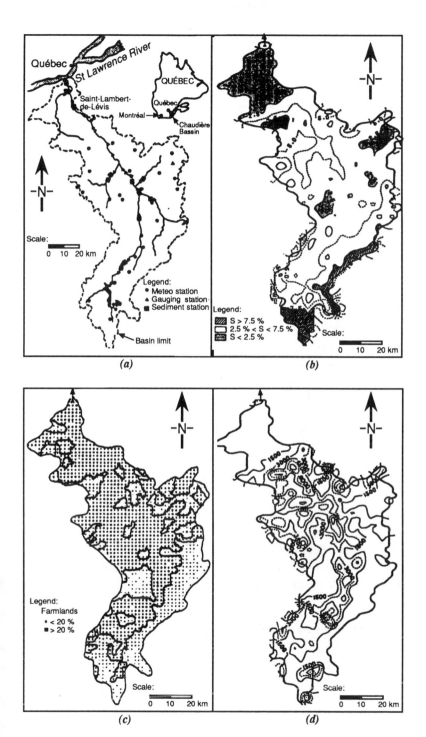

(a)

Québec
St Lawrence River
Saint-Lambert-
de-Lévis

QUÉBEC
Québec
Montréal→
Chaudière
Bassin

Scale:
0 10 20 km

Legend:
● Meteo station
▲ Gauging station
■ Sediment station

— Basin limit

(b)

-N-

Legend:
▨ S > 7.5 %
☐ 2.5 % < S < 7.5 %
▓ S < 2.5 %

Scale:
0 10 20 km

(c)

-N-

Legend:
Farmlands
· < 20 %
■ > 20 %

Scale:
0 10 20 km

(d)

-N-

1500
1500
1500
1500
1500
1500

Scale:
0 10 20 km

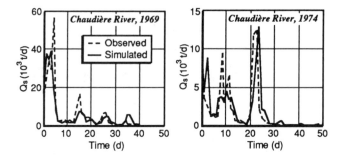

Figure CS11.3.2. Daily sediment load simulation of the Chaudière River (after Frenette and Julien, 1986b)

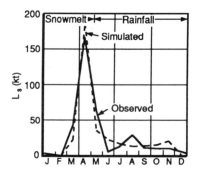

Figure CS11.3.3. Monthly sediment load distribution of the Chaudière River (after Julien, 1982)

*Problem 11.1

Compute the average sediment concentration C_{ppm} in an alluvial canal using the methods of Engelund and Hansen, Ackers and White, and Yang. The bed material (specific gravity 2.65) has the following particle size distribution:

Fraction diameter (mm)	Geometric mean (mm)	Fraction by weight (Δp_i)
0.062–0.125	0.088	0.04
0.125–0.25	0.177	0.23
0.25–0.50	0.354	0.37
0.50–1.0	0.707	0.27
1.0–2.0	1.414	0.09

The canal carries a discharge of 105 m³/s with a water temperature of 15°C. The channel has a slope of 0.00027, an alluvial bed width of 46 m, a flow depth of 2.32 m, and a sideslope of 2:1.

Answer: Engelund–Hansen, 356 ppm; Ackers–White, 866 ppm; Yang, 140 ppm.

**Problem 11.2

The Conca de Tremp watershed covers 43.1 km² in Spain. The elevation ranges from 530 to 1,460 m above sea level; the climate is typically Mediterranean with 690 mm of mean annual precipitation and a 12.5°C mean annual temperature. The Mediterranean forest has been depleted and the region has been intensively farmed for centuries. With reference to the following upland erosion map in tons/hectare × year (after Julien and Gonzalez del Tanago, 1991):

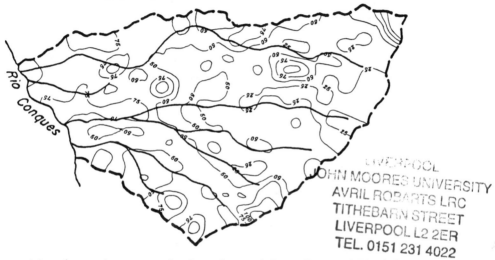

(a) estimate the gross upland erosion and the sediment yield of the watershed;
(b) how does the erosion rate compare with (i) the geological erosion rate, 0.1 tons/acre × year (1 acre = 0.40468 hectare); (ii) accelerated erosion rates for pasture, 5 tons/acre × year; and (iii) the erosion rate of urban development, 50 tons/acre × year?

*Problem 11.3

Consider sediment transport in the Elkhorn River, Waterloo, Nebraska, given the total drainage area of 6,900 mi². The flow-duration curve and the sediment-rating curve are detailed in the following tabulations:

Flow-duration curve		Sediment-rating curve	
% time exceeded	Discharge (ft³/s)	Discharge (ft³/s)	Suspended load (tons/day)
0.05	37,000	280	250
0.3	15,000	500	600
1.0	9,000	800	1,000
3.25	4,500	1,150	3,000
10	2,100	1,800	8,000
20	1,200	2,300	18,000
30	880	4,200	40,000
40	710	6,400	90,000
50	600	8,000	300,000
60	510	10,000	500,000
70	425		
80	345		
90	260		
96.75	180		

Calculate (a) the mean annual suspended sediment load using the flow-duration/sediment-rating-curve method; and (b) the sediment yield per square mile.

Answer: (a) 4.9×10^6 tons/year; (b) 710 tons/mi² × year.

**Computer Problem 11.1

The Niobrara River, Nebraska, carries a significant volume of sand. The bed-material size has a geometric mean of 0.283 mm and standard deviation $\sigma_g = 1.6$, $d_{35} = 0.233$ mm, $d_{50} = 0.277$ m, $d_{65} = 0.335$ mm, and $d_{90} = 0.53$ mm. The detailed sieve analysis is given in the following tabulation:

Sieve opening (mm)	% Finer
0.062	0.05
0.125	4.2
0.250	40.0
0.05	89.0
1.0	96.5
2.0	98.0
4.0	99.0

The stage–discharge relationship for the Niobrara River is shown in the adjacent figure at flow depths ranging from 0.7 to 1.3 ft and unit discharges from 1.7 to 5 ft²/s. The washload consists of particles finer than 0.125 mm. Given the channel width of 110 ft, slope of 0.00129, and water temperature of 60°F:

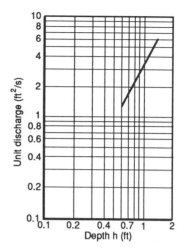

(a) calculate the unit sediment discharge by size fractions using three methods given in this chapter;
(b) plot the results on the given sediment-rating curve in Figure 11.6; and
(c) compare the equations with the observed unit sediment discharge in Figure 11.6.

*Computer Problem 11.2

Consider the channel reach analyzed in Computer Problems 3.1 and 8.2. Select one appropriate bed sediment discharge relationship to calculate the bed sediment discharge in tons/m × day by size fractions for the sediment size distribution considered in Computer Problem 8.2b. Plot the total sediment transport capacity along the 25-km reach; discuss the methods, assumptions, and results.

12

Reservoir sedimentation

As natural streams enter reservoirs, the stream flow depth increases and the flow velocity decreases. This reduces the sediment transport capacity of the stream and causes settling. Sediments carried into a reservoir may deposit throughout its full length, thus raising the bed elevation in time and causing aggradation. The pattern of deposition generally begins with a deltaic formation in the reservoir headwater area. Finer sediment particles may be transported by density currents down to the dam, thus completing the depositional pattern. Figure 12.1 depicts a typical reservoir sedimentation pattern. Aggradation in the upstream channel may occur over long distances above the reservoir because of the reduction in velocity and sediment-transporting capacity of the stream in this reach.

Reservoir sedimentation is a complex process that varies with watershed sediment production, rate of transportation, and mode of deposition. Reservoir sedimentation depends on the river regime, flood frequencies, reservoir geometry and operation, flocculation potential, sediment consolidation, density currents, and possible land use changes over the

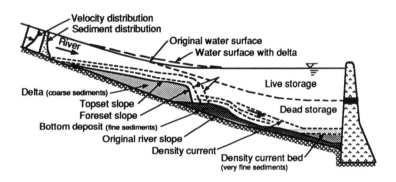

Figure 12.1. Typical reservoir sedimentation pattern
(after Frenette and Julien, 1986a)

242

life expectancy of the reservoir. In the analysis of reservoir sedimentation, storage losses in terms of live and dead storage, trap efficiency, control measures, and the operation of the reservoir must be considered given: the inflow hydrograph, the sediment inflow, the sediment characteristics, the reservoir configuration, the regional geography, and land use. The life expectancy of reservoirs indicates the time at which a reservoir is expected to become entirely filled with sediment. Its evaluation represents a challenge to planners, since the sediment sources arise from various geological formations. Cutting and burning of brushland and forest, overgrazed grasslands, and natural hazards including landslides, typhoons, volcanoes, and changes in land use are likely to occur during the expected life of the reservoir. Once the incoming sediment load has been determined from Section 12.1, the analysis of backwater profiles in Section 12.2 is combined with an analysis of sediment transport capacity to calculate the aggradation rate and the trap efficiency in Section 12.3. Sediment deposits consolidate over time (Section 12.4), and the life expectancy of a reservoir can be estimated from the information given in Section 12.5. An analysis of density currents (Section 12.6) is sometimes conducive to sediment management techniques such as flushing. Reservoir sedimentation surveys and sediment control measures are briefly covered in Sections 12.7 and 12.8, respectively.

12.1 Incoming sediment load

The incoming sediment load must be measured at appropriate gaging stations over several years before construction. Flow and sediment measurements define the sediment-rating curve described in Section 11.3. The annual suspended sediment yield is obtained from the sediment-rating curve and the flow-duration curve, as shown in Section 11.4. Difficulties in measuring bedload call for the use of bedload formulas, such as the Meyer-Peter and Müller, Einstein and Brown, and Duboys formulas. Armored beds in mountainous streams are common, and the composition of the bedload is different from that of the bed material. In some cases consolidated clay is found beneath a thin layer of alluvium covering the bed that is washed out at high flows. The calculated load is larger than the measured load when the transport capacity of the stream is greater than the availability of sediments. The ratio of bedload to suspended load can sometimes be estimated from Figure 10.4.

When washload is dominant, mathematical models can be used to predict soil losses by overland flow (Case Study 11.3). Watershed models using the universal soil-loss equation, among others, can be used when sufficient data are available from topographic, land use, and agricultural maps, Geographic Information Systems, aerial photographs, and field surveys. Sediment yield estimates based on physical characteristics of watershed are extremely valuable because the rate of sediment transport can then be predicted for alternative watershed conditions.

12.2   Reservoir hydraulics

As a stream enters a reservoir, the flow depth increases and the velocity and friction slope decrease, as generally described by M-1 backwater curves (Fig. E3.7.1). Most streams can be analyzed by one-dimensional approximations of the equations of conservation of mass and momentum. The resulting backwater equation has been derived in Example 3.7. The average flow depth h varies with downstream distance x as a function of the bed slope S_0, the friction slope S_f, and the Froude number $Fr = V/\sqrt{gh}$ given the mean flow velocity V and the gravitational acceleration g:

$$\frac{dh}{dx} = \frac{S_0 - S_f}{1 - Fr^2} \tag{E3.7.1}$$

Equation (E3.7.1) can be solved numerically from the point of maximum depth at the downstream end. The reach length Δx corresponding to a small change in water depth Δh is calculated by solving Equation (E3.7.1) or (E3.7.3):

$$\Delta x = \frac{\Delta h}{S_0 - S_f}\left(1 - \frac{V^2}{gh}\right) \tag{12.1}$$

At the upstream end, the normal flow depth corresponds to $\Delta h \to 0$, $\Delta x \to \infty$, and $S_f = S_0$, while in the reservoir $Fr^2 \to 0$, $S_f \to 0$, and $\Delta x = \Delta h/S_0$. Solution to Equation (12.1) is sought while satisfying the continuity relationship $Q = VhW$ given the channel width W. The resistance equation is also required in terms of hydraulic radius and friction slope, as compiled in Table 6.1.

12.3   Trap efficiency and aggradation

Owing to the continuity of sediment, part of the total load deposits on the channel bed as the sediment transport capacity decreases in the downstream direction. A delta is formed and the reservoir is gradually filled with sediments. With reference to the sediment continuity relationship [Eq. (10.3)] without sediment source ($\dot{C} = 0$):

$$\frac{\partial C}{\partial t} + \frac{\partial \hat{q}_{tx}}{\partial x} + \frac{\partial \hat{q}_{ty}}{\partial y} + \frac{\partial \hat{q}_{tz}}{\partial z} = 0 \tag{12.2}$$

where the mass fluxes $\hat{q}_{tx}$, $\hat{q}_{ty}$, and $\hat{q}_{tz}$ were defined in Equation (10.4).

Assuming a steady supply of sediment ($\partial C/\partial t = 0$), Equation (12.2) for one-dimensional flow ($\partial \hat{q}_{ty}/\partial y = 0$) reduces to

$$\frac{\partial \hat{q}_{tx}}{\partial x} + \frac{\partial \hat{q}_{tz}}{\partial z} = 0 \tag{12.3}$$

It is further assumed that the diffusive and mixing fluxes from Equation (10.4) are small compared with the advective fluxes in a reservoir. Considering settling to be the dominant advective flux in the vertical direction, $v_z = -\omega$, one obtains from Equations (10.4) and (12.3)

$$\frac{\partial v_x C}{\partial x} - \frac{\partial \omega C}{\partial z} = 0 \tag{12.4}$$

A practical approximation may be obtained for gradually varied flow ($\partial v_x/\partial x \to 0$), constant fall velocity ω, and $\partial C/\partial z = -C/h$; thus,

$$v_x \frac{\partial C}{\partial x} + \frac{\omega C}{h} = 0 \tag{12.5}$$

The solution for grain sizes of a given fraction i (constant fall velocity) at a constant unit discharge $q = Vh$, given $v_x = V$, is a function of the upstream sediment concentration C_{0i} of fraction i at $x = 0$:

$$C_i = C_{0i} e^{-X\omega_i/hV} \tag{12.6}$$

This shows that the concentration left in suspension is negligible ($C_i/C_{0i} = 0.01$) at a distance X_{C_i}:

$$X_{C_i} = 4.6 \frac{hV}{\omega_i} \tag{12.7}$$

The percentage of sediment fraction i that settles within a given distance X defines the trap efficiency T_{E_i} as

$$T_{E_i} = \frac{C_{0i} - C_i}{C_{0i}} = 1 - e^{-X\omega_i/hV} \tag{12.8} \blacklozenge$$

Without resuspension, 99% of the sediment in suspension settles within a distance $X_{C_i} = 4.6hV/\omega_i$. When calculating the trap efficiency of silt and clay particles, careful consideration must be given to density currents (Section 12.6) and possible flocculation, in which case the flocculated settling velocity ω_{fi} from Section 5.4.3 must be used instead of ω_i.

The settling sediment flux in the z direction for a given size fraction i causes a change in bed surface elevation z. Given the porosity $p_0 = \forall_v/\forall_t = 1 - C_v$ of the bed material, the integrated form of Equation (12.3) over the depth h with q_{txi} as the unit sediment discharge from Equation (11.2) in L^2/T is

$$T_{E_i} \frac{\partial q_{txi}}{\partial x} + (1 - p_0) \frac{\partial z_i}{\partial t} = 0 \qquad (12.9a)$$

or

$$\frac{\partial z_i}{\partial t} = -\frac{T_{E_i}}{(1 - p_0)} \frac{\partial q_{txi}}{\partial x} \qquad (12.9b) \blacklozenge$$

Values of porosity p_0 depend on the specific weight of sediment deposits covered in Section 12.4. For distances separating successive cross sections ΔX larger than X_C, the trap efficiency is unity and aggradation responds directly to changes in the sediment transport capacity of the stream. For ΔX smaller than X_C, only part of the sediment load in suspension will settle within the given reach. The sediment load at the downstream end will then exceed the sediment transport capacity of the stream.

12.4   Dry specific weight of sediment deposits

The conversion of the incoming weight of sediment to volume necessitates knowledge of the average dry specific weight of a mixture γ_{md}, defined in Chapter 2 as the dry weight of sediment per unit total volume including voids. For material coarser than 0.1 mm the specific dry weight of the mixture remains practically constant around $\gamma_{md} = 14.75$ kN/m^3, or 93 lb/ft^3. The corresponding dry mass density of the mixture ρ_{md} is given by $\rho_{md} = \gamma_{md}/g = 1{,}500$ kg/m^3, or 2.9 slugs/ft^3. The porosity p_0 of sand material is then obtained from $p_0 = 1 - \gamma_{md}/\gamma_s = 0.43$. The volumetric sediment concentration $C_v = 1 - p_0$ and the void ratio $e = p_0/(1 - p_0)$.

Under pressure, the dry specific weight of finer sediment fractions varies in time due to the consolidation of the material and exposure to the air. After T_c years the dry specific weight of a mixture γ_{mdT} increases as a function of time from the initial dry specific weight γ_{md1} after $T_c = 1$ year according to

$$\gamma_{mdT} = \gamma_{md1} + K \log T_c \qquad (12.10)$$

Values of the initial dry specific weight γ_{md1} and consolidation factor K in lb/ft^3 are compiled in Table 12.1. Assuming continuous uniform

Table 12.1. *Dry specific weight of sediment deposits mixture* (γ_{mdl} *and* K, *lb/ft*^{3})

	Lane and Koelzer						Trask					
	Sand		Silt		Clay		Sand		Silt		Clay	
	γ_{mdl}	K	γ_{mdl}	K	γ_{mdl}	K	γ_{mdl}	K	γ_{mdl}	K	γ_{mdl}	K
Sediment always submerged or nearly submerged	93	0	65	5.7	30	16.0	88	0	67	5.7	13	16.0
Normally moderate reservoir drawdown	93	0	74	2.7	46	10.7	88	0	76	2.7	—	10.7
Normally considerable reservoir drawdown	93	0	79	1.0	60	6.0	88	0	81	1.0	—	6.0
Reservoir normally empty	93	0	82	0.0	78	0.0	88	0	84	0.0	—	0.0

Note: 62.4 lb/ft^3 = 9,810 N/m^3.

settling during a period of T_c years, the depth-averaged dry specific weight of the mixture after T_c years, $\bar{\gamma}_{mdT}$, is given by Miller's formula:

$$\bar{\gamma}_{mdT} = \gamma_{mdl} + 0.43K\left[\frac{(T_c \ln T_c)}{T_c - 1} - 1\right] \tag{12.11}$$

where

$\bar{\gamma}_{mdT}$ = depth-averaged dry specific weight after T_c years (lb/ft^3)

γ_{mdl} = initial dry specific weight (lb/ft^3)

K = consolidation factor (lb/ft^3)

T_c = consolidation time (years)

In the case of heterogeneous sediment mixtures, the specific weight of a mixture is calculated using weight-averaged values for each size fraction, as shown in Example 12.1.

Example 12.1 Application to density of sediment deposits. The source of sediment entering a large reservoir contains 20% sand, 65% silt, and 15% clay.

(a) Determine the dry specific weight of the mixture after 1 year and after 100 years, considering nearly submerged conditions; refer to the following tabulation:

Size class	Fraction Δp_i	γ_{md1} (lb/ft^3)	$\Delta p_i \times \gamma_{md1}$ (lb/ft^3)	K (lb/ft^3)	$\bar{\gamma}_{md100}$ (lb/ft^3)	$\Delta p_i \times \bar{\gamma}_{md100}$ (lb/ft^3)
Sand	0.20	93	18.6	0	93	18.6
Silt	0.65	65	42.2	5.7	74	48.1
Clay	0.15	30	4.5	16.0	55	8.2
Total	1.00		$\bar{\gamma}_{md1} = 65.3$ lb/ft^3 after 1 year			$\bar{\gamma}_{md100} = 75$ lb/ft^3 after 100 years

(b) Calculate the porosity after 1 year and after 100 years:

$$p_{0\ (1\ \text{year})} = 1 - \frac{\bar{\gamma}_{md1}}{\gamma_s} = 1 - \frac{65.3}{2.65 \times 62.4} = 0.60$$

$$p_{0\ (100\ \text{years})} = 1 - \frac{\bar{\gamma}_{md100}}{\gamma_s} = 1 - \frac{75}{2.65 \times 62.4} = 0.55$$

12.5   Life expectancy of reservoirs

The life expectancy of a reservoir is the expected time at which the reservoir will be completely filled with sediments. Its determination requires knowledge of the storage capacity or volume of the reservoir Ψ_R, the mean annual incoming total sediment discharge Q_t in weight per year, the sediment size distribution, the trap efficiency of the reservoir T_E, and the dry specific weight of sediment deposits γ_{md}. After transforming the incoming mean annual sediment discharge by size fraction i into volume of sediment trapped in the reservoir, one obtains the life expectancy T_R by

$$T_R = \frac{\Psi_R \gamma_{mdT}}{\sum_i T_{E_i} \Delta p_i Q_{ti}} \qquad (12.12) \blacklozenge$$

The life expectancy represents an average duration on which the economic feasibility of the reservoir can be based. The accuracy of life expectancy calculations depends on the annual sediment discharge. Mean annual values are useful for long-term estimates.

The analysis of extreme events is also important when the life expectancy is short and the variability in mean annual sediment yield is large (e.g., in arid areas). The probability that one or several severe events may fill the reservoir before T_R must then be considered. For instance, the risk of an extreme flood occurring in the next five years and its impact on the life expectancy must be calculated. Such an extreme event could severely deplete the live storage and restrict further use of the reservoir for hydropower generation and storage of flood water.

In reservoirs having a reduced capacity–inflow ratio, it is important to consider aspects of the incoming sediment load and the settling capacity of fine particles: (1) What are the likely changes in land use of the watershed in terms of potential increase in sediment production from upland areas during the life expectancy of the reservoir? (2) What is the possible impact of an extreme event such as a 1,000-year flood on the storage capacity and operation of the reservoir? (3) What is the potential effect of flocculation on changes in the trap efficiency of the reservoir?

12.6   Density currents

A density current may be defined as the movement under gravity of fluids of slightly different density. For sediment-laden flows in reservoirs, the underflow is caused by sediment-laden river water with a higher specific weight than the clear water in the reservoir.

In general, engineering problems associated with density currents are concerned with the passage of sediment through the reservoir and with the transfer of sediment from the live storage to the dead storage. Ultimately, a complete solution would therefore require the following: (a) a determination of initial velocity, depth, and sediment transport rate of density currents from known characteristics of the reservoir and the river inflow; (2) an estimate of the amount of interfacial mixing resulting in the decreasing sediment-carrying capacity; (3) a determination of whether the density current will be stable throughout the length of the reservoir (with the objective of providing sluicing devices to discharge sediment downstream from the dam).

Density currents consist chiefly of particles in suspension of less than 20 μm in diameter. Stokes' law gives a settling rate for particles of this diameter of approximately 0.001 ft/s. Thus, transverse turbulent fluctuations of the order of 1% in a current having a mean velocity of only 0.1 ft/s would be sufficient to keep such particles in suspension. Sediment particles found in most density currents are commonly referred to as *washload,* originating from erosion on the land slopes of the drainage area rather than from the streambed. In a river, the concentration of such material is practically constant from bed to surface and is relatively independent of major changes in flow conditions that occur at the plunge point of the reservoir.

The relation between the mass density difference $\Delta\rho$, or the specific weight difference $\Delta\gamma$, and the volumetric concentration is described by

$$\frac{\Delta\rho}{\rho} = \frac{\Delta\gamma}{\gamma} = C_v(G-1) \qquad (12.13)$$

By analogy with open-channel flow, the average velocity V_d of the density current of thickness h_d given the bed slope S_0 of the reservoir is a function of the densimetric friction factor f_d:

$$V_d = \sqrt{\frac{8\Delta\rho g h_d S_0}{\rho f_d (1 + \alpha_d)}} \tag{12.14}$$

The density current unit discharge corresponds to $q_d = V_d h_d$ and the sediment flux by volume q_{tv} is calculated by $q_{tv} = C_v q_d$, because the sediment concentration is relatively uniform.

For the case of laminar flow, calculations involve the densimetric Reynolds number $\mathrm{Re}_d = V_d h_d / \nu < 1{,}000$, and the corresponding value of $\alpha_d = 0.64$. The densimetric friction factor f_d can be expressed as a function of the Reynolds number:

$$f_d(1 + \alpha_d) = 1.64 f_d = \frac{58}{\mathrm{Re}_d}$$

In the case of turbulent flow, $\mathrm{Re}_d = V_d h_d / \nu > 1{,}000$, the densimetric friction factor can be approximated by $f_d \simeq 0.01$ and $\alpha_d \simeq 0.5$.

Any attempt to refine the analysis of turbulent flows is hindered by the fact that as the degree of turbulence increases, the interface becomes increasingly difficult to define, due to mixing and resulting vertical density variations. The interface of a density current at very low velocities is smooth and distinct, and consists of a sharp discontinuity of density across which the velocity variation is continuous. As the relative velocity between the two layers is increased, waves are formed at the interface, and at a certain critical velocity the mixing process begins by the periodic breaking of the interfacial waves.

Various criteria for determining the flow conditions at which mixing begins have been proposed. Keulegan (1949) defined a mixing stability parameter ϑ from viscous and gravity forces as

$$\vartheta = \left(\frac{\nu g \Delta\rho}{V_d^3 \rho}\right)^{1/3} = \left(\frac{1}{\mathrm{Fr}_d^2 \mathrm{Re}_d}\right)^{1/3} \tag{12.15}$$

in which the densimetric Froude number $\mathrm{Fr}_d = V_d / \sqrt{(\Delta\rho/\rho) g h_d}$ and the densimetric Reynolds number $\mathrm{Re}_d = V_d h_d / \nu$.

In laminar flow, mixing occurs when the flow reaches critical densimetric depth ($\mathrm{Fr}_d = 1$) or $\vartheta_c = 1/\mathrm{Re}_d^{1/3}$. The average experimental value of ϑ_c for turbulent flow is $\vartheta_c = 0.18$. The flow is stable when $\vartheta > \vartheta_c$ and mixing occurs when $\vartheta < \vartheta_c$. Example 12.2 provides an application of this method to the density currents observed in Lake Mead.

Example 12.2 Application of density currents. Surveys of Lake Mead indicate the existence of underflows when the density difference between the current and the surrounding water is of the order of $\Delta\rho/\rho = 0.0005$. The average bed slope of the reservoir is about 5 ft/mi, or $S_0 = 0.00095$. If measurements indicate a density current depth of approximately 15 ft, the magnitude of the velocity can be obtained from Equation (12.14) assuming $f_d = 0.010$ and $\alpha_d = 0.5$ for turbulent flow:

$$V_d = \sqrt{\frac{8(0.0005)32.2 \text{ ft}}{0.010(1.5) \text{ s}^2}} \times \sqrt{15 \text{ ft}(0.00095)} = 0.35 \frac{\text{ft}}{\text{s}}$$

The density current Reynolds number is therefore

$$\text{Re}_d = \frac{V_d h_d}{\nu} = 5 \times 10^5$$

The computed velocity of 0.35 ft/s (approximately 6 mi/d) is consistent with field measurements. The stability of the density current is calculated from Equation (12.15):

$$\vartheta = \left[\frac{1.2 \times 10^{-5} \text{ ft}^2}{\text{s}} \times \frac{32.2 \text{ ft}}{\text{s}^2} \times \frac{0.0005 \text{ s}^3}{(0.35)^3 \text{ ft}^3} \right]^{1/3} = 0.0165$$

Mixing will occur because $\vartheta < 0.18$ for turbulent flow.

The rate of sediment transport by the density current can also be estimated. From Equation (12.13), $C_v = 0.0003$ or $C = 803$ mg/l, and if the average width W_d of the density current is taken as 500 ft, the discharge represented by the density current is $Q_d = V h_d W_d = 2,625$ ft^3/s and that of sediment $Q_{sd} = Q_d C_v = 0.78$ ft^3/s, or 130 lb/s, which is approximately equivalent to a sediment transport rate of 5,600 tons/d.

12.7   Reservoir sedimentation surveys

Reservoir surveys obtained after the closure of a dam provide useful information on the deposition pattern, trap efficiency, and density of deposited sediment. This information may be necessary for efficient reservoir operation. Before a new survey is undertaken, the original and/or preceding ones should be studied. Subsequently a decision should be made as to which technique will be used for the forthcoming survey. There are three basic survey techniques:

1. For filled reservoirs: sonic equipment measuring primarily the reservoir bottom elevation, and terrestrial or aerial photogrammetric survey measuring the water surface area at a given stage

2. For empty reservoirs: aerial photogrammetric survey of the reservoir topography
3. For partly empty reservoirs: a combination of the above

In addition it is essential to sample the deposited sediment in order to determine its specific mass density and particle size distribution.

Electronic surveying equipment can be used quite effectively. Airborne or satellite photogrammetric methods have proved to be economically attractive; they require a minimum of ground control, and with stereoscopic equipment a contour map can easily be drawn.

Sonic sounders or fathometers, operating from a boat crossing the reservoir, detect the reservoir bottom continuously. A sonic sounder consists of a sensor unit containing the transmitting and receiving transducer and a recording unit, which continuously records the water depth on a chart. Sonic sounders are sometimes capable of delineating the differences in densities of submerged deposits.

Sediment samplers, such as gravity-core, piston-core, and spud-rod samplers, are used to take undisturbed volumetric samples of deposited sediments. Radioactive gamma probes determine the specific weight of the deposit in situ without removing a sample.

12.8   Control measures

Because of the large number of variables involved in reservoir sedimentation problems, no single control measure can be suggested. To simplify the discussion, control measures can be grouped into categories: watershed control, inflow control, and deposition control.

Control of the watershed, if at all possible, may be the most effective sediment control measure, since it reduces the sediment production at the source. Such a control measure is certainly feasible with small watersheds; however, it may be an expensive long-term undertaking with larger ones. Proper soil conservation practices provide an effective means of reducing erosion. Increasing the vegetative cover of a watershed is very important for sheet, gully, and channel erosion control.

The control of sediment inflow into a reservoir can be achieved by proper watershed management supplemented with sediment-retarding structures throughout the watershed. Stream-channel improvement and stabilization should be considered, including the building of settling (or debris) basins, sabo dams, or off-channel reservoirs and the utilization of existing or new bypass canals or vegetation screens (since dense plant growth reduces the flow velocity and causes deposition).

Control of deposition begins with the proper design of a reservoir, particularly the position and operation of outlets, spillways, and possibly sluice gates drawing sediment-laden density currents directly. After deposition has occurred, various methods exist for removing sediments. Methods such as dredging, flushing, and sluicing must be minimized because they are expensive and often ineffective. In reservoirs with a low capacity–inflow ratio, flushing for several days during the flood season is sometimes sufficient to wash out several years of accumulated sediments. Flushing reservoirs with a large capacity–inflow ratio is generally counterproductive. Two case studies illustrate various aspects of reservoir sedimentation.

Case Study 12.1   Tarbela Reservoir, Pakistan. The Tarbela Dam is located on the Indus River near Islamabad in Pakistan. It is the world's largest rock- and earth-filled dam, being 2.74 km long and 143 m high. Built at a cost of about 1.5 billion U.S. dollars, it has the capacity to generate 3,750 MW of hydropower. The reservoir spans more than 80 km upstream of the dam (Fig. CS12.1.1) with a total storage capacity of 11.3 million acre-feet, or 13.9 km³. The upstream watershed covers 171,000 km², of which only 6% receives monsoon rain; the annual precipitation on the remainder of the watershed does not exceed 10 cm.

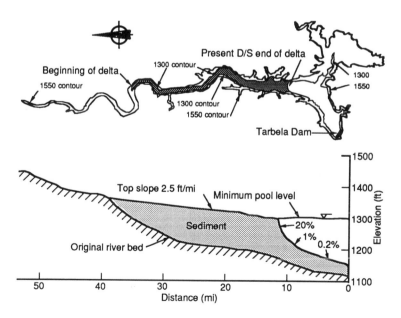

Figure CS12.1.1. Tarbela Dam

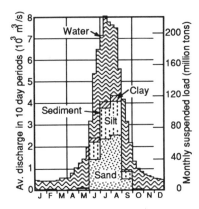

Figure CS12.1.2. Discharge and sediment load, Indus River

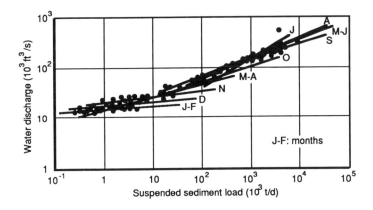

Figure CS12.1.3. Sediment-rating curve, Indus River

Surface runoff and incoming sediment load depend primarily on melting snow (Figs. CS12.1.2 and CS12.1.3). The mean annual flow based on a 115-year record is 78.9 km^3/year, which is about five to six times the reservoir storage capacity. The average annual sediment load is of the order of 200 million tons/year. With a sediment size distribution of 60% sand, 33% silt, and 7% clay, a dry specific weight of about 1,350 kg/m^3 is estimated. With a trap efficiency estimated at 90%, the depletion in storage capacity due to sedimentation is about 0.135 km^3/year, from which the life expectancy is about 100 years.

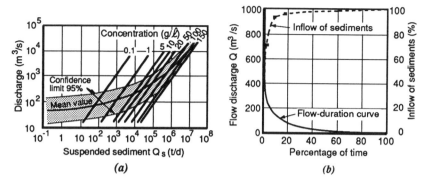

Figure CS12.2.1. Caine River: (a) sediment-rating curve;
(b) flow-duration curve (after Frenette and Julien, 1986a)

Case Study 12.2   Molineros Reservoir, Bolivia. The projected Molineros Reservoir is located on the Caine River in the Bolivian Andes. Large sediment loads in the river have postponed the otherwise viable project since 1972. In 1983, the proposed project included a 200-m-high dam with hydropower production reaching 132 MW. The 45-km-long reservoir in a narrow gorge has a live storage capacity of 2.98 km^3.

The steep watershed is sparsely covered with ground vegetation on very erodible sandstone. About 80% of the 630-mm mean annual precipitation on the 9,530-km^2 watershed occurs during the wet season, which extends from December to March. The mean annual flow is 47.2 m^3/s, with possible flood flows reaching 3,750 and 4,150 m^3/s at a period of return of 1,000 and 10,000 years, respectively.

The sediment-rating curve and the flow duration curve are shown in Figure CS12.2.1; sediment concentrations as high as 150,000 mg/l have been measured. The analysis of annual suspended sediment load gives 92×10^6 tons/year, as shown in Figure CS12.2.2. Assuming that about 15% of the total load is bedload, a mean annual sediment load of 108×10^6 tons/year is estimated. Given that the sediment size distribution is about 25% sand, 65% silt, and 10% clay, the dry specific weight of reservoir deposits ranges from 1,100 to 1,300 kg/m^3. The trap efficiency of this large reservoir approaches 100% for all size fractions. Considering a reservoir capacity of close to 3 km^3, the life expectancy varies between 19 and 35 years depending on various alternatives selected. Figure CS12.2.2 shows that, in this case, the inflow of sediment during a 100-year flood is equivalent to about four times the mean annual sediment inflow.

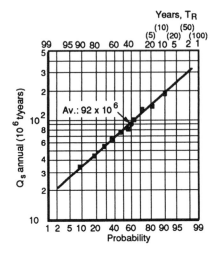

Figure CS12.2.2. Annual suspended load versus probability
(after Frenette and Julien, 1986a)

**Problem 12.1

From the data given in Case Study 12.2 on the Molineros Reservoir Project,

    (a) determine the trap efficiency and the specific weight of sediment deposits after 10 years;

    (b) use the flow-duration/sediment-rating-curve method to estimate the annual sediment load;

    (c) calculate the life expectancy of the reservoir; and

    (d) examine the impact of a 1,000-year flood in the next five years on the life expectancy of the reservoir.

*Computer Problem 12.1

Consider the channel reach of Computer Problem 3.1. Calculate the monthly changes in bed elevation profiles (aggradation–degradation) for bedload transport of the uniform 1-mm sand size from Computer Problems 8.2a and 9.1. Estimate the life expectancy of the unit width reservoir.

    Repeat the calculations for total load given the sand size distribution considered in Computer Problems 8.2b and 11.2. Estimate the life expectancy of the reservoir and compare the results with those above.

Appendix: Einstein's sediment transport method

Einstein's method (1950) is herein presented using the amended original notation. Einstein's bed sediment discharge function gives the rate at which flow of any magnitude in a given channel transports the individual sediment sizes found in the bed material. His equations are extremely valuable in many studies for determining the time change in bed material when each size moves at its own rate. For each size D_s of the bed material, the bed load discharge is given as

$$i_B q_B \qquad (A.1)$$

the suspended sediment discharge is given by

$$i_s q_s \qquad (A.2)$$

the total bed sediment discharge is

$$i_T q_T = i_s q_s + i_B q_B \qquad (A.3)$$

and, finally,

$$Q_T = \Sigma i_T q_T \qquad (A.4)$$

where i_T, i_s, and i_B are the fractions of the total, suspended, and contact bed sediment discharges, q_T, q_s and q_B, for a given grain size D_s. The term Q_T is the total bed sediment discharge. The suspended sediment discharge is related to the bed sediment discharge because there is a continuous exchange of particles between the two modes of transport.

With suspended sediment discharge related to the bed sediment discharge, Equation (A.3) becomes

$$i_T q_T = i_B q_B (1 + P_E I_1 + I_2) \qquad (A.5) \blacklozenge$$

where

$$i_B q_B = \phi_* i_B \gamma_s \left(\frac{\rho}{\rho_s - \rho} \frac{1}{gD_s^3} \right)^{-1/2} \qquad (A.6)$$

257

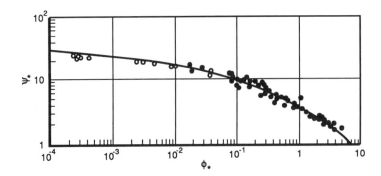

Figure A.1. Einstein's $\phi_* - \psi_*$ bedload function (Einstein, 1950)

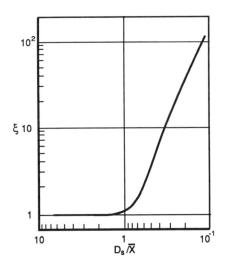

Figure A.2. Hiding factor (Einstein, 1950)

and

γ_s = the unit weight of sediment

ρ = the density of water

ρ_s = the density of sediment

g = gravitational acceleration

ϕ_* = dimensionless sediment transport function $f(\psi_*)$ given in Figure A.1

$$\psi_* = \xi Y (\log 10.6/B_x)^2 \psi \tag{A.7}$$

$$\psi = \left(\frac{\rho_s - \rho}{\rho}\right) \frac{D_s}{R_b' S_f} \tag{A.8}$$

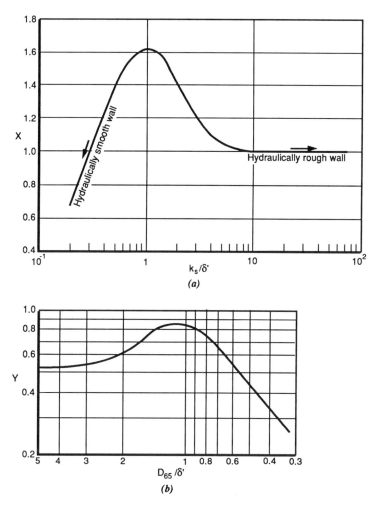

Figure A.3. (a) Einstein's multiplication factor X in the logarithmic velocity equations; (b) pressure correction (Einstein, 1950)

ξ = a correction factor given as a function of $D_s/\bar{X}$ in Figure A.2

$$\bar{X} = 0.77\Delta \quad \text{if} \quad \Delta/\delta' > 1.8 \tag{A.9a}$$

$$\bar{X} = 1.39\delta' \quad \text{if} \quad \Delta/\delta' < 1.8 \tag{A.9b}$$

Δ = the apparent roughness of the bed, k_s/X;

X = a correction factor in the logarithmic velocity distribution, given as a function of k_s/δ' in Figure A.3

$$\delta' = 11.6\nu/V'_*$$ (A.10)

$V/V'_* =$ Einstein's velocity distribution

$$= 5.75 \log(30.2y_0/\Delta)$$ (A.11)

$V'_* =$ the shear velocity due to grain roughness

$$V'_* = \sqrt{gR'_b S_f}$$ (A.12)

$R'_b =$ the hydraulic radius of the bed due to grain roughness

$$= R_b - R''_b$$

$R''_b =$ the hydraulic radius of the bed due to channel irregularities

$S_f =$ the slope of the energy grade line normally taken as the slope of the water surface;

$Y =$ another correction term, given as a function of D_{65}/δ' in Figure A.3b

$B_x = \log(10.6\bar{X}/\Delta)$.

The preceding equations are used to compute the fraction i_B of the load. The other terms in Equation (A.5) are

$$P_E = 2.3 \log 30.2y_0/\Delta$$ (A.13)

I_1 and I_2 are integrals of Einstein's form of the suspended sediment Equation (A.2):

$$I_1 = 0.216 \frac{E^{Z-1}}{(1-E)^Z} \int_E^1 \left[\frac{1-y}{y}\right]^Z dy$$ (A.14)

$$I_2 = 0.216 \frac{E^{Z-1}}{(1-E)^Z} \int_E^1 \left[\frac{1-y}{y}\right]^Z \ln y \, dy$$ (A.15)

where

$Z = \omega/0.4V'_*$

$\omega =$ the fall velocity of the particle of size D_s

$E =$ the ratio of bed layer thickness to flow depth, a/y_0

$y_0 =$ depth of flow

$a =$ the thickness of the bed layer, $2D_{65}$

The two integrals I_1 and I_2 are given, respectively, in Figures A.4 and A.5 as a function of Z and E.

In the preceding calculations for the total load, the shear velocity is based on the hydraulic radius of the bed due to grain roughness R'_b. Its

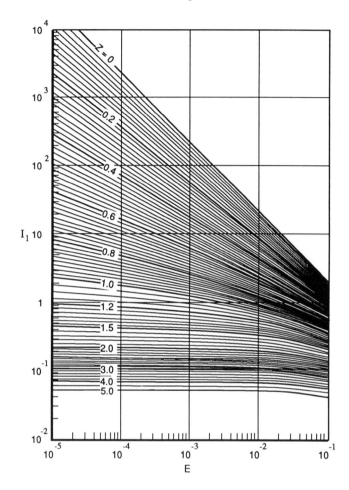

Figure A.4. Integral I_1 in terms of E and Z (Einstein, 1950)

computation can be explained as follows. Total resistance to flow is composed of two parts, surface drag and form drag. The transmission of shear to the boundary is accompanied by a transformation of flow energy into turbulence. The part of energy corresponding to grain roughness is transformed into turbulence, which stays at least for a short time in the immediate vicinity of the grains and has a great effect on the bedload motion. The part of the energy corresponding to the form resistance is transformed into turbulence at the interface between wake and free stream flow, or at a considerable distance from the grains. This energy does not contribute to the bedload motion of the particles and may be largely neglected in the sediment transportation.

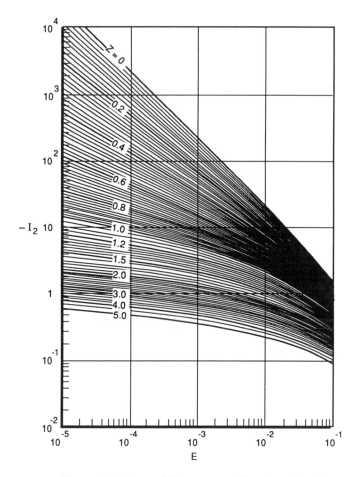

Figure A.5. Integral I_2 in terms of E and Z (Einstein, 1950)

Einstein's equation for mean flow velocity V in terms of V'_* is

$$V/V'_* = 5.75 \log(12.26R'_b/\Delta) \tag{A.16a}$$

or

$$V/V'_* = 5.75 \log(12.26R'_b X/k_s) \tag{A.16b}$$

Furthermore, Einstein suggested that

$$V/V''_* = \mathfrak{F}[\psi'] \tag{A.17}$$

where

$$\psi' = \left(\frac{\rho_s - \rho}{\rho}\right)\frac{D_{35}}{R'_b S_f} \tag{A.18}$$

The relation for Equation (A.18) is given in Figure A.6. The procedure to follow in computing R'_b depends on the information available. If mean

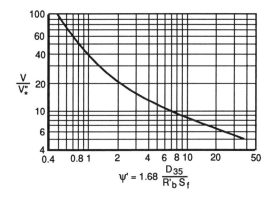

Figure A.6. V/V_*'' versus ψ' (Einstein, 1950)

velocity V, friction slope S_f, hydraulic radius R_b, and bed-material size are known, then R_b is computed by trial and error using Equation (A.16) and Figure A.6.

The procedure for computing total bed sediment discharge in terms of different size fractions of the bed material is as follows:

1. calculate ψ^* using Equation (A.7) for each size fraction;
2. find ϕ_* from Figure A.1 for each size fraction;
3. calculate $i_B q_B$ for each size fraction using Equation (A.6);
4. sum up the q_B across the flow to obtain $i_B Q_B$; and
5. sum up the size fractions to obtain Q_B.

For the suspended sediment discharge,

6. calculate z for each size fraction using Equation (A.16);
7. calculate $E = 2D_s/y_0$ for each fraction;
8. determine I_1 and I_2 for each fraction from Figures A.4 and A.5;
9. calculate P_E using Equation (A.13);
10. compute the suspended discharge from $i_B q_B (P_E I_1 + I_2)$; and
11. sum up all the q_B and i_B to obtain the total suspended discharge Q_{ss}.

Thus, to get the total bed sediment discharge,

12. add the results of Steps 5 and 11.

A sample problem showing the calculation of the total bed sediment discharge using Einstein's procedure is presented in Example A.1.

Example A.1   Total bed sediment discharge calculation from Einstein's method. A test reach, representative of the Big Sand Creek near Green-

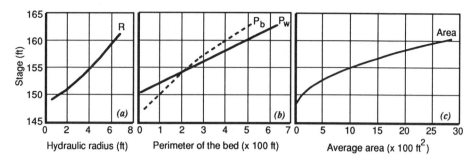

Figure A.7. Description of the average cross section (Einstein, 1950)

wood, Mississippi, was used by Einstein (1950) as an example of applying his bedload function. For simplicity, the effects due to bank friction are neglected here; the reader can refer to the original example for the construction of the representative cross section. The channel slope of this cross section was determined to be $S_f = 0.00105$. The relation between cross-sectional area, hydraulic radius, and wetted perimeter of the representative cross section and stage is given in Figure A.7. In the case of this wide and shallow channel, the wetted perimeter is assumed to equal the surface width. The averaged values of the four bed-material samples are given in Table A.1. Ninety-six percent of the bed material is between 0.147 and 0.589 mm, which is divided into four size fractions.

Table A.1. *Bed sediment information for sample problem*

Grain size distribution (mm)	Average grain size			Settling velocity	
	mm	ft	i_B (%)	cm/sec	ft/s
$D_s > 0.60$	—	—	2.4	—	—
$0.60 > D_s > 0.42$	0.50	0.00162	17.8	6.2	0.205
$0.42 > D_s > 0.30$	0.36	0.00115	40.2	4.5	0.148
$0.30 > D_s > 0.21$	0.25	0.00081	32.0	3.2	0.106
$0.21 > D_s > 0.15$	0.18	0.00058	5.8	2.0	0.067
$0.15 > D_s$	—	—	1.8	—	—
$D_{16} = 0.24$ mm					
$D_{35} = 0.29$ mm	$\sigma_g = 1.35$				
$D_{50} = 0.34$ mm					
$D_{65} = 0.37$ mm	Gr $= 1.35$				
$D_{84} = 0.44$ mm					

Table A.2. *Hydraulic calculations for sample problem in applying the Einstein procedure*

R_b'	V_*'	δ'	κ_s/δ'	X	Δ	V	ψ'	V/V_*''	V_*''	R_b''
0.5	0.129	0.00095	1.21	1.59	0.00072	2.92	2.98	16.8	0.17	0.86
1.0	0.184	0.00067	1.72	1.46	0.00079	4.44	1.49	27.0	0.16	0.76
2.0	0.259	0.00047	2.44	1.27	0.00090	6.63	0.75	51.0	0.13	0.50
3.0	0.318	0.00039	2.95	1.18	0.00097	8.40	0.50	87.0	0.10	0.30
4.0	0.368	0.00033	3.50	1.14	0.00102	9.92	0.37	150.0	0.07	0.14
5.0	0.412	0.00030	3.84	1.10	0.00104	11.30	0.30	240.0	0.05	0.07
6.0	0.450	0.00027	4.26	1.08	0.00107	12.58	0.25	370.0	0.03	0.03

R_b	y_0	Z_0	A	P_b	Q	$\bar{X}$	Y	B_x	$(B/B_x)^2$	P_E
1.36	1.36	150.2	140	103	409	0.00132	0.84	1.29	0.63	10.97
1.76	1.76	150.9	240	136	1,065	0.00093	0.68	1.12	0.85	11.10
2.50	2.50	152.1	425	170	2,820	0.00069	0.56	0.91	1.27	11.30
3.30	3.30	153.3	640	194	5,380	0.00076	0.55	0.91	1.27	11.50
4.14	4.14	154.9	970	234	9,620	0.00079	0.54	0.91	1.27	11.70
5.07	5.07	156.9	1,465	289	16,550	0.00080	0.54	0.91	1.27	11.90
6.03	6.03	159.5	2,400	398	30,220	0.00082	0.54	0.91	1.27	12.04

Definitions: R_b', bed hydraulic radius due to grain roughness, ft; V_*', shear velocity due to grain roughness, ft/s $= \sqrt{gR_b'S_f}$; δ', thickness of the laminar sublayer, ft $= 11.62\nu/V_*'$; κ_s, roughness diameter, ft $= D_{65}$; X, correction factor in the logarithmic velocity distribution, given in Figure A.3a; Δ, apparent roughness diameter, ft $= \kappa_s/X$; V, average flow velocity, ft/s $= 5.75V_*'\log(12.27R_b'/\Delta)$; ψ', intensity of shear on representative particles, equal to $[(\rho_s-\rho)/\rho](D_{35}/R_b'S_f)$; V/V_*'', velocity ratio, given in Figure A.6; V_*'', shear velocity due to form roughness, ft/s; R_b'', bed hydraulic radius due to form roughness, ft $= V_*''^2/gS$; R_b, bed hydraulic radius, ft $= R$, the total hydraulic radius if there is no additional friction $= R_b'+R_b''$; y_0, average flow depth, ft $= R$ for wide, shallow streams; Z_0, stage, ft from Figure A.7; A, cross-sectional area, ft^2; P_b, bed wetter perimeter, ft; Q, flow discharge $= AV$; $\bar{X}$, characteristic distance, from Equation (A.9); Y, pressure correction term, given in Figure A.3b; B_x, coefficient $= \log(10.6\bar{X}/\Delta)$; B, coefficient $= \log 10.6$; P_E, Einstein's transport parameter $= 2.303\log(30.2y_0/\Delta)$.
Source: After Einstein (1950).

The sediment transport calculations are made for the individual size fraction that has a representative grain size equal to the geometric mean grain diameter of each fraction. The water viscosity is $\nu = 1.0 \times 10^{-5}$ ft^2/s and the specific gravity of the sediment is 2.65.

Important hydraulic parameters are given in Table A.2. The bed sediment transport is then calculated for each grain fraction of the bed material at each given flow depth. It is convenient to summarize the calculations in the form of tables. The procedure is given in Table A.3.

Table A.3. Bed sediment load calculations for sample problem in applying the Einstein procedure

D	i_B	R'_b	Ψ	$D/\bar{X}$	ξ	Ψ_*	ϕ_*	$i_B q_B$	$i_B Q_B$	$\Sigma i_B Q_B$	$10^3 E$	Z	I_1	$-I_2$	$P_E I_1 + I_2 + 1$	$i_T q_T$	$i_T Q_T$	$\Sigma i_T Q_T$
.00162	.178	0.5	5.08	1.23	1.08	2.90	1.90	0.0267	119.00	400	2.38	3.78	0.078	0.44	1.42	0.03800	168	670
		1.0	2.54	1.74	1.00	1.73	4.00	0.0561	330.00	1,335	1.84	2.65	0.131	0.74	1.71	0.09580	561	3,938
		2.0	1.27	2.35	1.00	0.90	8.20	0.1150	845.00	3,771	1.30	1.88	0.240	1.27	2.44	0.28100	2,050	30,500
		3.0	0.85	2.16	1.00	0.60	12.80	0.1800	1,510.00	6,496	0.98	1.53	0.385	2.01	3.44	0.61700	5,170	113,000
		4.0	0.63	2.05	1.00	0.43	18.00	0.2530	2,560.00	10,745	0.78	1.33	0.560	2.80	4.75	1.20000	12,100	324,000
		5.0	0.51	2.03	1.00	0.35	22.50	0.3160	3,950.00	16,333	0.63	1.18	0.810	3.85	6.78	2.13000	26,500	800,000
		6.0	0.42	1.98	1.00	0.29	27.00	0.3800	6,350.00	27,142	0.54	1.08	1.090	4.90	9.20	3.48000	59,800	1,940,000
.00115	.402	0.5	3.38	0.82	1.36	2.44	2.45	0.0471	210.00		1.69	2.88	0.117	0.68	1.60	0.07540	335	
		1.0	1.69	1.16	1.10	1.27	5.50	0.1060	623.00		1.31	2.02	0.210	1.19	2.14	0.22700	1,330	
		2.0	0.85	1.57	1.01	0.61	12.60	0.2420	1,780.00		0.92	1.44	0.450	2.33	3.76	0.91000	6,660	
		3.0	0.56	1.44	1.04	0.41	19.00	0.3640	3,050.00		0.70	1.17	0.830	3.85	6.73	2.44000	20,400	
		4.0	0.42	1.37	1.05	0.30	26.00	0.5000	5,050.00		0.56	1.01	1.370	5.70	11.30	5.65000	57,100	
		5.0	0.34	1.35	1.05	0.25	31.50	0.6040	7,540.00		0.45	0.90	2.120	8.10	17.20	10.40000	129,000	
		6.0	0.28	1.32	1.05	0.20	39.00	0.7490	12,900.00		0.38	0.83	2.950	10.50	26.00	19.60000	335,000	
.00081	.320	0.5	2.54	0.61	2.25	3.03	1.75	0.0155	69.00		1.19	1.94	0.230	1.29	2.23	0.03450	153	
		1.0	1.27	0.87	1.26	1.09	6.80	0.0600	353.00		0.92	1.36	0.520	2.60	4.16	0.25000	1,460	
		2.0	0.63	1.17	1.10	0.49	15.80	0.1390	1,020.00		0.65	0.97	1.530	6.10	12.20	1.70000	12,500	
		3.0	0.42	1.08	1.12	0.33	23.50	0.2070	1,730.00		0.49	0.79	3.350	11.00	28.70	5.95000	49,700	
		4.0	0.32	1.04	1.15	0.25	31.50	0.2790	2,820.00		0.39	0.68	6.200	17.50	56.00	15.60000	157,000	
		5.0	0.25	1.01	1.17	0.20	39.50	0.3490	4,360.00		0.32	0.61	9.800	25.50	92.00	32.00000	397,000	
		6.0	0.21	0.99	1.19	0.17	46.00	0.4060	6,980.00		0.27	0.55	15.000	36.00	146.00	59.50000	1,020,000	

.00057	.058	0.5	1.80	0.43	5.40	5.15	0.58	0.00056	2.49	0.85	1.21	0.720	3.35	5.55	0.00312	14
		1.0	0.90	0.61	2.28	1.39	5.10	0.00500	29.40	0.65	0.86	2.440	8.10	20.00	0.10000	587
		2.0	0.45	0.83	1.37	0.44	17.50	0.01710	126.00	0.46	0.61	8.400	21.50	74.40	1.26000	9,350
		3.0	0.30	0.76	1.52	0.32	25.00	0.02460	206.00	0.35	0.49	19.300	41.00	183.00	4.50000	37,600
		4.0	0.22	0.72	1.60	0.25	31.50	0.03100	313.00	0.28	0.43	32.000	63.00	312.00	9.68000	97,800
		5.0	0.18	0.71	1.65	0.20	39.50	0.03870	483.00	0.23	0.38	51.000	91.00	516.00	20.00000	248,000
		6.0	0.15	0.70	1.70	0.18	43.50	0.04260	752.00	0.19	0.35	70.000	122.00	722.00	30.80000	526,000

Definitions: D, representative grain size, ft, given in Table A.1; i_B, fraction of bed material given in Table A.1; R'_b, bed hydraulic radius due to grain roughness, ft, given in Table A.2; ψ, intensity of shear on a particle $= [(\rho_s - \rho)/\rho](D/R'_b S_f)$; $D/\bar{X}$, dimensionless ratio, $\bar{X}$ given in Table A.2; ξ, hiding factor, given in Figure A.2; ψ_*, intensity of shear on individual grain size $= \xi Y(B/B_x)^2 \psi$ [values of Y and $(B/B_x)^2$ are given in Table A.2]; ϕ_*, intensity of sediment transport for individual grain from Figure A.1; $i_B q_B$, bedload discharge per unit width for a size fraction, lb/s × ft $= i_B \phi_* \rho_s (gD)^{3/2} \sqrt{(\rho_s/\rho) - 1}$; $i_B Q_B$, bedload discharge for a size fraction for entire cross section, tons/d $= 43.2 W i_B q_B$, $W = P_b$, given in Table A.2 (1 ton $= 2{,}000$ lb); $\sum i_B Q_B$, total bedload discharge for all size fractions for entire cross section, tons/d; E, ratio of bed layer thickness to water depth $= 2D/y_0$ (for values of y_0, see Table A.2); Z, exponent for concentration distribution, $\omega/(0.4 U'_*)$ (for values of ω and U'_* see Tables A.1 and A.2); I_1, integral, given in Figure A.4; $-I_2$, integral, given in Figure 1.5; $P_E I_1 + I_2 + 1$, factor between bedload and total load, using P_E in Table A.2; $i_T q_T$, bed-material load per unit width of stream for a size fraction, lb/s × ft $= i_B q_B (P_E I_1 + I_2 + 1)$; $i_T Q_T$, bed-material load for a size fraction of entire cross section, tons/d $= 43.2 P_b i_B q_B$ (P_b given in Table A.2); $\sum i_T Q_T$, total bed-material load for all size fractions, tons/d.
Source: Einstein (1950).

Bibliography

Ackers, P., and W. R. White. Sediment transport: New approach and analysis. *J. Hyd. Div. ASCE, 99,* no. HY11 (1973): 2041-60.

Adriaanse, M. De ruwheid van de Bergsche Maas bij hoge afvoeren. Rijkswaterstaat, RIZA, Nota 86.19. Dordrecht, 1986.

Alonso, C. V., W. H. Neibling, and G. R. Foster. Estimating sediment transport capacity in watershed modeling. *Trans. ASAE, 24,* no. 5 (1991): 1211-20.

American Society of Civil Engineers. *Sedimentation Engineering.* ASCE Manuals and Reports on Engineering Practice no. 54. New York, 1975.

Athaullah, M. Prediction of bedforms in erodible channels. Ph.D. thesis, Colorado State University, 1968.

Bagnold, R. A. Experiments on a gravity-free dispersion of large solid spheres in a Newtonian fluid under shear. *Proc. Roy. Soc. Lond. A225* (1954): 49-63.

Flow of cohesionless grains in fluids. *Phil. Trans. Roy. Soc. Lond., 249,* no. 964 (1956): 235-97.

An approach to the sediment transport problem from general physics. Professional Paper 422-I. U.S. Geological Survey, Washington, D.C.: 1966.

Barekyan, A. S. Discharge of channel forming sediments and elements of sand waves. *Soviet Hydrol. Selected Papers* (1962): 128-30.

Beverage, J. P., and J. K. Culbertson. Hyperconcentrations of suspended sediment. *J. Hyd. Div. ASCE, 90,* no. HY6 (1964): 117-28.

Bingham, E. C. *Fluidity and Plasticity.* New York: McGraw-Hill, 1922.

Bird, R. B., W. E. Stewart, and E. N. Lightfoot. *Transport Phenomena.* New York: Wiley, 1960.

Bishop, A. A., D. B. Simons, and E. V. Richardson. Total bed material transport. *J. Hyd. Div. ASCE, 91,* no. HY2 (1965): 175-91.

Bogardi, J. *Sediment Transport in Alluvial Streams.* Budapest: Akademiai Kiado, 1974.

Borland, W. M. Reservoir sedimentation. In *River Mechanics,* ed. H. W. Shen. Ft. Collins, Colo.: Water Resources Pub., 1971.

Borland, W. M., and C. R. Miller. Distribution of sediment in large reservoirs. *J. Hyd. Div. ASCE, 84,* no. HY2 (1958): 1-18.

Boyce, R. Sediment routing and sediment-delivery ratios. In *Present and Prospective Technology for Predicting Sediment Yields and Sources,* USDA-ARS-S-40, 1975, pp. 61-5.

269

Bray, D. I. Flow resistance in gravel-bed rivers. In *Gravel-Bed Rivers, Fluvial Processes, Engineering and Management*, ed. R. D. Hey, J. C. Bathurst, and C. R. Thorne. New York: Wiley, 1982, pp. 109–37.

Bridgman, P. W. *Dimensional Analysis*. New Haven, Conn.: Yale University Press, 1922.

Brooks, N. H. Mechanics of streams with movable beds of fine sediment. *Trans. ASCE, 123* (1958): 525–49.

Brown, C. B. Sediment transportation. In *Engineering Hydraulics*, ed. H. Rouse. New York: Wiley, 1950, pp. 769–857.

Brownlie, W. R. Prediction of flow depth and sediment discharge in open-channels. Report no. KH-R-43A. Pasadena: California Institute of Technology, W. M. Keck Laboratory, 1981.

Buckingham, E. On physical similar systems: Illustration of the use of dimensionless equations. *Phys. Rev., 4,* no. 4 (1914).

Cartens, M. R. Accelerated motion of a spherical particle. *Trans. Amer. Geophys. Union, 33,* no. 5 (1952).

Chabert, J., and J. L. Chauvin. Formation de dunes et de rides dans les modèles fluviaux. *Bull. Cen. Rech. ess. Chatou,* no. 4 (1963).

Chang, F. M., D. B. Simons, and E. V. Richardson. Total bed material discharge in alluvial channels. Water Supply Paper 1498-I. Washington, D.C.: U.S. Geological Survey, 1965.

Chien, N. The present status of research on sediment transport. *Trans. ASCE, 121* (1956): 833–68.

Chow, V. T. *Open Channel Hydraulics*. New York: McGraw-Hill, 1964.

Colby, B. R. Discontinuous rating curves for Pigeon Roost and Cuffawa Creeks in northern Mississippi. ARS 41-36. Washington, D.C.: U.S. Department of Agriculture, 1960.

   Discharge of sands and mean velocity relationships in sand-bed streams. Professional Paper 462-A. Washington, D.C.: U.S. Geological Survey, 1964a.

   Practical computations of bed material discharge. *J. Hyd. Eng., 90,* no. HY2 (1964b): 217–46.

Colby, B. R., and C. H. Hembree. Computations of total sediment discharge Niobrara River near Cody, Nebraska. Water Supply Paper 1357. Washington, D.C.: U.S. Geological Survey, 1955.

Coleman, N. L. Velocity profiles with suspended sediment. *J. Hyd. Res., 19,* no. 3 (1981): 211–29.

   Effects of suspended sediment on the open-channel velocity distribution. *Water Resources Res., 22,* no. 10 (1986): 1377–84.

Coles, D. E. The law of the wake in the turbulent boundary layer. *J. Fluid Mech., 1* (1956): 191–226.

Dawdy, D. R. Studies of flow in alluvial channels, depth, height and discharge relation for alluvial streams. Water Supply Paper 1798c. Washington, D.C.: U.S. Geological Survey, 1961.

Duboys, M. P. Etudes du régime du Rhone et de l'action exercée par les eaux sur un lit à fond de graviers indéfiniment affouillable. *Ann. Ponts et Chaussées,* ser. 5, *18* (1879): 141–95.

Edwards, T. K., and E. D. Glysson. Field methods for measurement of fluvial sediment. Open-File Report 86-531. Reston, Va: U.S. Geological Survey, 1988.

Egiazaroff, I. V. Calculation of non-uniform sediment concentration. *J. Hyd. Div. ASCE, 91,* no. HY4 (1965): 225-48.

Einstein, H. A. Formulas for the transportation of bed load. *Trans. ASCE, 107* (1942): 561-73.

The bed load function for sediment transport in open channel flows. Technical Bulletin no. 1026. Washington, D.C.: U.S. Department of Agriculture, Soil Conservation Service, 1950.

Sedimentation, Part II: River sedimentation. In *Handbook of Applied Hydrology,* ed. V. T. Chow. New York: McGraw-Hill, 1964, sec. 17.

Einstein, H. A., and N. Chien. Effects of heavy sediment concentration near the bed on velocity and sediment distribution. MRD Series no. 8. Berkeley: University of California: Institute of Engineering Research; Omaha, Neb.: U.S. Army Engineering Div., Missouri River, Corps of Engineers, 1955.

Emmett, W. W. A field calibration of the sediment trapping characteristics of the Helley–Smith bedload sampler. Open File Report 79-411. Denver, Colo.: U.S. Geological Survey, 1979.

Engelund, F. Hydraulic resistance of alluvial streams. *J. Hyd. Div. ASCE, 92,* no. HY2 (1966): 315-26.

Engelund, F., and E. Hansen. *A Monograph on Sediment Transport to Alluvial Streams.* Copenhagen: Teknik Vorlag, 1967.

Etcheverry, B. A. Irrigation practice and engineering: The conveyance of water. *Trans. ASCE, 11* (1916).

Fischer, H. B., E. J. List, R. C. Koh, J. Imberger, and N. H. Brooks. *Mixing in Inland and Coastal Waters.* New York: Academic Press, 1979.

Fortier, A. *Mécanique des Suspensions.* Monographie de Mécanique des Fluides et Thermique. Paris: Masson, 1967.

Fortier, S., and F. C. Scobey. Permissible canal velocities. *Trans. ASCE, 89,* paper no. 1588 (1926): 940-84.

Frenette, M., and P. Y. Julien. Computer modeling of soil erosion and sediment yield from large watersheds. *Proc. 27th Annual Convention of the Institute of Engineers,* Peshawar, Pakistan, 1985.

Advances in predicting reservoir sedimentation. General lecture, Third International Symposium on River Sedimentation, ISRS-111, Jackson, Mississippi, March 31 to April 4, 1986a, pp. 26-46.

LAVSED I: A model for predicting suspended load in northern streams. *Can. J. Civ. Eng., 13,* no. 2 (1986b): 150-61.

Computer modeling of soil erosion and sediment yield from large watersheds. *Int. J. Sed. Res., 1* (1987): 39-68.

Frenette, M., J. C. Souriac, and J. P. Tournier. Modélisation de l'alluvionnement de la retenue de Péligre, Haiti. *Proc. 14th International Congress on Large Dams,* Rio de Janeiro, May 3-7, 1982, pp. 93-115.

Frenette, M., J. P. Tournier, and T. J. Nzakimuena. Cas historique de sédimentation du barrage Péligre, Haiti. *Can. J. Civ. Eng., 9,* no. 2 (1982): 206-23.

Garde, R. J., and K. G. Ranga Raju. *Mechanics of Sediment Transportation and Alluvial Stream Problems,* 2d ed. New York: Wiley, 1985.

Gessler, J. Beginning and ceasing of sediment motion. In *River Mechanics,* ed. H. W. Shen. Littleton, Colo.: Water Resources Pub., 1971, chap. 7.

Gilbert, G. K. The transport of debris by running water. Professional Paper 86. Washington, D.C.: U.S. Geological Survey, 1914.

Goldstein, S. The steady flow of viscous fluid past a fixed spherical obstacle at small Reynolds number. *Proc. Roy. Soc. Lond., 123A* (1929): 225–35.

Govier, G. W., C. A. Shook, and E. O. Lilge. The properties of water suspension of finely subdivided magnetite, galena, and ferrosilicon. *Trans. Can. Inst. Mining Met., 60* (1957): 147–54.

Graf, W. H. *Hydraulics of Sediment Transport.* New York: McGraw-Hill, 1971.

Guy, H. P., and V. W. Norman. USGS techniques of water resources investigations. In *Fluid Methods for Measurement of Fluvial Sediments.* Washington, D.C.: U.S. Government Printing Office, 1970, book 3, chap. C-2.

Guy, H. P., D. B. Simons, and E. V. Richardson. Summary of alluvial channel data from flume environments, 1956–1961. Professional Paper 462-I. Washington, D.C.: U.S. Geological Survey, 1966.

Happel, J., and H. Brenner. *Low Reynolds Number Hydrodynamics.* Englewood Cliffs, N.J.: Prentice-Hall, 1965.

Hartley, D. M., and P. Y. Julien. Boundary shear stress induced by raindrop impact. *J. Hyd. Res. IAHR, 30,* no. 3 (1992): 351–9.

Henderson, F. M. *Open Channel Flow.* New York: Macmillan, 1966.

Highway Research Board. Tentative design procedure for riprap-lined channels. Report no. 108. Washington, D.C.: National Academy of Sciences, National Cooperative Highway Research Program, 1970.

Holtorff, F. Steady bed material transport in alluvial channels. *J. Hyd. Div. ASCE, 109,* no. 3 (1983): 368–84.

Hunsaker, J. C., and B. G. Rightmire. *Engineering Applications of Fluid Mechanics.* New York: McGraw-Hill, 1947.

Interagency Committee. The design of improved types of suspended sediment samplers. Report no. 6. Iowa City: Subcommittee on Sedimentation, Federal Interagency River Basin Committee, Hydrology Laboratory of the Iowa Institute of Hydraulic Research, 1952.

Jenson, V. J. Viscous flow round a sphere at low Reynolds numbers. *Proc. Roy. Soc. Lond., A249* (1959): 346–66.

Julien, P. Erosion de bassin et apport solide en suspension dans les cours d'eau nordiques. M.S. thesis, Laval University, 1979.

Prédiction d'apport solide pluvial et nival dans les cours d'eau nordiques à partir du ruissellement superficiel. Ph.D. dissertation, Laval University, 1982.

Downstream hydraulic geometry of noncohesive alluvial channels. *Proc. International Conference on River Regime,* Wallingford, England, May 1988, pp. 9–16.

Study of bedform geometry in large rivers. Delft Hydraulics, Report Q1386. Emmeloord, 1992.

Julien, P. Y., and M. Frenette. Predicting washload from rainfall and snowmelt runoff. *Proc. Third Internation Symposium on River Sedimentation,* ISRS-111, Jackson, Mississippi, March 31 to April 4, 1986a, pp. 1259–65.

LAVSED II: A model for predicting suspended load in northern streams. *Can. J. Civ. Eng., 13,* no. 2 (1986b): 162–70.

Macroscale analysis of upland erosion. *Hyd. Sci. J. IAHS, 32,* no. 3 (1987): 347–58.

Julien, P. Y., and M. Gonzalez del Tanago. Spatially-varied soil erosion under different climates. *Hyd. Sci. J. IAHS, 36,* no. 6 (1991): 511–24.

Julien, P. Y., and Y. Q. Lan. Rheology of hyperconcentrations. *J. Hyd. Eng. ASCE, 115,* no. 3 (1991): 346–53.

Julien, P. Y., Y. Q. Lan, and G. Berthault. Experiments on stratification of heterogeneous sand mixtures. *Bull. Soc. Géol. France, 164,* no. 5 (1993): 649–60.

Julien, P. Y., and D. B. Simons. Sediment transport capacity of overland flow. *Trans. ASAE, 28,* no. 3 (1985a): 755–62.

Sediment transport capacity equations for rainfall erosion. *Proc. Fourth International Hydrology Symposium,* Fort Collins, Colo., July 15–17, 1985b, pp. 988–1002.

Kalinske, A. A. Criteria for determining sand transport by surface creep and saltation. *Trans. AGU, 23,* part 2 (1942): 639–43.

Movement of sediment as bed load in rivers. *Trans. AGU, 28,* no. 4 (1947): 615–20.

Kamphuis, J. W. Determination of sand roughness for fixed beds. *J. Hyd. Res. IAHR, 12,* no. 2 (1974): 193–203.

Karim, F., and J. F. Kennedy. Computer-based predictors for sediment discharge and friction factor of alluvial streams. Report no. 242. Iowa Institute of Hydraulic Research, University of Iowa, 1983.

Keulegan, G. H. Interfacial instability and mixing in stratified flows. *J. Res. Nat. Bur. Stand., 43,* no. 487, RP 2040 (1949).

Kikkawa, H., and T. Ishikawa. Total load of bed materials in open channels. *J. Hyd. Div. ASCE, 104,* no. HY7 (1978): 1045–60.

Kilinc, M. Y. Mechanics of soil erosion from overland flow generated by simulated rainfall. Ph.D. dissertation, Colorado State University, 1972.

Klaassen, G. J., J. S. Ribberink, and J. C. C. de Ruiter. On sediment transport of mixtures. Pub. 394. Delft Hydraulics, Emmeloord, 1988.

Lacey, G. Stable channels in alluvium. *Proc. Inst. Civ. Eng., 229* (1929).

Lane, E. W. Progress report on studies on the design of stable channels by the Bureau of Reclamation. *Proc. ASCE,* no. 280 (1953).

Langhaar, H. L. *Dimensional Analysis and Theory of Models.* New York: Wiley, 1951.

Liu, H. K. Mechanics of sediment-ripple formation. *J. Hyd. Div. ASCE, 183,* no. HY2 (1957): 1–23.

Lowe, J., III, and I. H. R. Fox. Sedimentation in Tarbela Reservoir. *Proc. 14th International Congress on Large Dams,* Rio de Janeiro, May 1982, pp. 317–40.

Mantz, P. A. Incipient transport of fine grains and flakes by fluids-extended Shields diagram. *J. Hyd. Eng., 103,* no. 6 (1977): 601–16.

Meyer-Peter, E., and R. Müller. Formulas for bed-load transport. *Proc. 2d Meeting IAHR,* Stockholm, 1948, pp. 39–64.

Migniot, C. Tassement et rhéologie des vases. *La Houille Blanche,* no. 1 (1989): 11–29.

Miller, C. R. Determination of the unit weight of sediment for use in sediment volume computations. Memorandum, U.S. Bureau of Reclamation, Department of the Interior, Denver, February 1953.

Mirtskhoulava, T. E. *Basic Physics and Mechanics of Channel Erosion,* transl. R. B. Zeidler. Leningrad: Gideometeoizdat, 1988.

Murphy, P. J., and E. J. Aguirre. Bed load or suspended load. *J. Hyd. Div. ASCE, 111,* no. 1 (1985): 93–107.

Nordin, C. F. A preliminary study of sediment transport parameters, Rio Puerco near Bernardo, New Mexico. Professional Paper 462-C. Washington, D.C.: U.S. Geological Survey, 1963.

Aspects of flow resistance and sediment transport: Rio Grande near Bernalillo, NM. Water Supply Paper 1498-H. Washington, D.C.: U.S. Geological Survey, 1964.

O'Brien, J. S., and P. Y. Julien. Physical properties and mechanics of hyperconcentrated sediment flows. *Proc. ASCE Specialty Conference on the Delineation of Landslide, Flashflood and Debris Flow Hazards,* Utah Water Research Lab, Series UWRL/G-85/03, 1985, pp. 260–79.

Laboratory analysis of mudflow properties. *J. Hyd. Eng. ASCE, 114,* no. 8 (1988): 877–87.

Oseen, C. Hydrodynamik. In *Akademische Verlags Gesellschaft.* Leipzig: AVG, 1927, chap. 10.

Parker, G. P. C. Klingeman, and D. G. McLean. Bedload and size distribution in paved gravel-bed streams. *J. Hyd. Eng., 108,* no. 4 (1982): 544–71.

Petersen, M. *River Engineering.* Englewood Cliffs, N.J.: Prentice-Hall, 1986.

Piest, R. F., and C. R. Miller. Sediment sources and sediment yields. In *Sedimentation Engineering,* ASCE Manual and Reports on Engineering Practice no. 54. New York, 1977, chaps. 4 and 5.

Qian, N. Reservoir sedimentation and slope stability, technical and environmental effects. General Report Q54, 14th International Congress on Large Dams, Rio de Janeiro, May 1982.

Qian, N., and Z. Wan. A critical review of the research on the hyperconcentrated flow in China. Beijing: International Research and Training Centre on Erosion and Sedimentation, 1986.

Ranga Raju, K. G., R. J. Garde, and R. C. Bhardwaj. Total load transport in alluvial channels. *J. Hyd. Div. ASCE, 107,* no. HY2 (1981): 179–92.

Raudkivi, A. J. *Loose Boundary Hydraulics,* 3d ed. New York: Pergamon Press, 1990.

Rayleigh, Lord. The principle of similitude. *Nature, 95* (1915).

Richardson, E. V., and P. Y. Julien. Bedforms, fine sediments, washload and sediment transport. Third International Symposium on River Sedimentation, ISRS-III, Jackson, Mississippi, March 31 to April 4, 1986, pp. 854–74.

Richardson, E. V., Simons, D. B., and P. Y. Julien. *Highways in the River Environment.* Design and training manual for U.S. Department of Transportation, Federal Highway Administration, publication no. FHWA-HI-90-016. Washington, D.C., 1990.

Richardson, E. V., Woo, H. S., and P. Y. Julien. Research trends in sedimentation. In *Proceedings of the Symposium on Megatrends in Hydraulic Engineering,* ed. M. L. Albertson and C. N. Papadakis. Fort Collins: Colorado State University, 1986, pp. 299–330.

Rouse, H. Modern conception of the mechanics of turbulence. *Trans. ASCE, 102* (1937): 463–543.

Rouse, H. (ed.). *Advanced Mechanics of Fluids.* New York: Wiley, 1959.

Rubey, W. Settling velocities of gravel, sand and silt particles. *Am. J. Sci., 25,* no. 148 (1933): 325–38.

Savage, S. B., and S. McKeown. Shear stresses developed during rapid shear of concentrated suspension of large spherical particles between concentric cylinders. *J. Fluid Mech., 137* (1983): 453–72.

Schlichting, H. *Boundary Layer Theory.* New York: McGraw-Hill, 1968.

Schoklitsch, A. Geschiebetreib und die geschiebefracht. Wasser Kraft und Wasser Wirtschaft. *Jgg. 39,* no. 4 (1934).

Schwab, G. O., R. K. Frevert, T. W. Edminster, and K. K. Barnes. *Soil and Water Conservation Engineering,* 3d ed. New York: Wiley, 1981.

Sedov, L. I. *Similarity and Dimensional Methods in Mechanics,* transl. M. Hold. New York: Academic Press, 1959.

Shen, H. W., and C. S. Hung. An engineering approach to total bed-material load by regression analysis. *Proc. Sedimentation Symposium,* ed. H. W. Shen. Berkeley, Calif.: Water Resources Pub., 1972, chap. 14.
    Remodified Einstein procedure for sediment load. *J. Hyd. Div. ASCE, 109,* no. 4 (1983): 565–78.

Shen, H. W., and P. Y. Julien. Erosion and sediment transport. In *Handbook of Hydrology,* ed. D. R. Maidment. New York: McGraw-Hill, 1993, chap. 12.

Shields, A. *Anwendung der Aehnlichkeitsmechanik und der Turbulenz Forschung auf die Geschiebebewegung.* Berlin: Mitteilungen der Preussische Versuchanstalt für Wasserbau und Schiffbau, 1936.

Simons, D. B. Theory and design of stable channels in alluvial material. Ph.D. dissertation, Colorado State University, 1957.

Simons, D. B., R. M. Li, and W. Fullerton. Theoretically derived sediment transport equations for Pima County, Arizona. Prepared for Pima County DOT and Flood Control District, Tucson, Ariz. Ft. Collins, Colo.: Simons, Li and Assoc., 1981.

Simons, D. B., and E. V. Richardson. Form of bed roughness in alluvial channels. *Trans. ASCE, 128* (1963): 284–323.
    Resistance to flow in alluvial channels. Professional Paper 422-J. Washington, D.C.: U.S. Geological Survey, 1966.

Simons, D. B., and F. Senturk. *Sediment Transport Technology.* Littleton, Colo.: Water Resources Pub., 1977, revised, 1992.

Singamsetti, S. R. Diffusion of sediment submerged jet. *J. Hyd. Div. ASCE, 109,* no. HY2 (1966).

Smerdon, E. T., and R. P. Beasley. Critical cohesive forces in cohesive soils. *Agr. Eng., 42* (1961): 26–9.

Stevens, M. A., and D. B. Simons. Stability analysis for coarse granular material on slopes. In *River Mechanics*. Littleton, Colo.: Water Resources Pub., 1971, chap. 17.

Streeter, V. L. *Fluid Mechanics*, 5th ed. New York: McGraw-Hill, 1971.

Thomas, W. A. Scour and deposition in rivers and reservoirs. Report no. 723-G2-L2470. Vicksburg, Miss.: U.S. Army Corps of Engineers, Hydrology Engineering Center, 1976.

Toffaleti, F. B. A procedure for computation of the total river and sand discharge and detailed distribution, bed to surface. Technical Report no. 5. Vicksburg, Miss.: U.S. Army Corps of Engineers, Committee on Channel Stabilization, 1968.

Tsujimoto, T. Fractional transport rate and fluvial sorting. *Proc. Grain Sorting Seminar,* Versuchanstalt für Wasserbau, Mitteilungen 117, ETH. Zurich, 1992, pp. 227–49.

U.S. Waterways Experiment Station. Studies of river bed materials and their movement with special reference to the lower Mississippi River. USWES, paper no. 17. Vicksburg, Miss., 1935.

van Rijn, L. C. Sediment transport, Part II: Suspended load transport. *J. Hyd. Div. ASCE, 110,* no. 11 (1984a): 1613–41.

    Sediment transport, Part III: Bedforms and alluvial roughness. *J. Hyd. Div. ASCE, 110,* no. 12 (1984b): 1733–54.

Vanoni, V. A. Transportation of suspended sediment by water. *Trans. ASCE, 3,* paper no. 2267 (1946): 67–133.

Vanoni, V. A., and N. H. Brooks. Laboratory studies of the roughness and suspended load of alluvial streams. Report E-68. Pasadena: California Institute of Technology, Sedimentation Laboratory, 1957.

Vanoni, V. A., N. H. Brooks, and J. F. Kennedy. Lecture notes on sediment transportation and channel stability. Report KH-RI. Pasadena: California Institute of Technology, W. M. Keck Laboratory, 1960.

Wargadalam, J. Hydraulic geometry equations of alluvial channels. Ph.D. dissertation, Colorado State University, 1993.

Wijbenga, J. H. A. Analyse prototype-metingen (niet-) permanente ruwheid. Report Q1302. Emmeloord: Delft Hydraulics, 1991.

Williams, D. T., and P. Y. Julien. On the selection of sediment transport equations. *J. Hyd. Eng. ASCE, 115,* no. 11 (1989): 1578–81.

Wilson, K. C. Bed-load transport at high shear stress. *J. Hyd. Div. ASCE, 92,* no. 6 (1966): 44–59.

Wischmeier, W. H. Upslope erosion analysis. In *Environmental Impact on Rivers,* ed. H. W. Shen. Littleton, Colo.: Water Resources Pub., 1972, chap. 15.

Wischmeier, W. H., and D. D. Smith. *Predicting Rainfall Erosion Losses: A Guide to Conservation Planning.* USDA Agr. Handbook 537. Washington, D.C.: U.S. Department of Agriculture, 1978.

Wiuff, R. Transport of suspended material in open and submerged streams. *J. Hyd. Div. ASCE, 111,* no. 5 (1985): 774–92.

Woo, H. S., and P. Y. Julien. Turbulent shear stress in heterogeneous sediment-laden flows. *J. Hyd. Eng. ASCE, 116,* no. 11 (1990): 1416–21.

Woo, H. S., Julien, P. Y., and E. V. Richardson. Washload and fine sediment load. *J. Hyd. Eng. ASCE, 112,* no. 6 (1986): 541–5.

Suspension of large concentrations of sands. *J. Hyd. Eng. ASCE, 114,* no. 8 (1988): 888–98.

Yalin, M. S. Geometrical properties of sand waves. *J. Hyd. Div. ASCE, 90,* no. HY5 (1964): 105–19.

Mechanics of Sediment Transport, 2d ed. London: Pergamon Press, 1977.

Yalin, M. S., and E. Karahan. Inception of sediment transport. *J. Hyd. Div. ASCE, 105,* no. HY11 (1979): 1433–43.

Yang, C. T. Incipient motion and sediment transport. *J. Hyd. Div. ASCE, 99,* no. HY10 (1973): 1679–1704.

Yih, C. S. *Fluid Mechanics.* Ann Arbor, Mich.: West River Press, 1979.

Young, R. A., and C. K. Mutchler. Soil movement on irregular slopes. *Water Resources Res., 5,* no. 5 (1969): 1084–5.

Yucel, O., and W. H. Graf. Bed load deposition in reservoirs. *Proc. 15th Congress of IAHR,* Istanbul, 1973.

Zingg, A. W. Degree and length of land slope as it affects soil loss by runoff. *Agricultural Engineering, 21,* no. 2 (1940): 59–64.

Index